Superplant:
Creating a Nimble Manufacturing Enterprise with Adaptive Planning Software

Shaun Snapp

Superplant: Creating a Nimble Manufacturing Enterprise with Adaptive Planning Software
Copyright © 2013 by SCM Focus Press

ALL RIGHTS RESERVED

For information about this title or to order other books and/or electronic media, contact the publisher:
SCM Focus Press
PO Box 29502 #9059
Las Vegas, NV 89126-9502
http://www.scmfocus.com/scmfocuspress
(408) 657-0249

ISBN: 978-1-939731-21-0

Printed in the United States of America

Cover and Interior design by: 1106 Design

Contents

CHAPTER 1: Introduction 1

CHAPTER 2: Understanding a Superplant Conceptually 29

CHAPTER 3: Multi-plant Planning 31

CHAPTER 4: Single Versus Multi-pass Planning 81

CHAPTER 5: Multi-source Planning 89

CHAPTER 6: Subcontracting Planning and Execution 95

CHAPTER 7: Combining All Three Superplant Functionalities 105

CHAPTER 8: The Superplant and the Integration Between ERP and the
External Planning System 113

CHAPTER 9: Superplant-enabled Capable-to-promise 121

CHAPTER 10: Conclusion 135

APPENDIX A: Labor Pools in Galaxy APS 139

APPENDIX B: Time Horizons in Galaxy APS 145

APPENDIX C: Prioritizing Internal Demand for Subcomponents over
External Demand 151

References 165

Vendor Acknowledgements and Profiles 169

Author Profile 171

Abbreviations 175

Links Listed in the Book by Chapter 177

CHAPTER 1

Introduction

This book addresses several production and supply planning software functionalities that are all related to the location-based adaptability of the supply chain planning application (multi-plant planning and subcontracting, and contract manufacturing planning). This adaptability is a historical weakness of both advanced planning applications as well as

ERP systems. Some of this functionality is rarely found in commercially-available applications, while other functionality is more commonly found but is difficult to implement. For more than a decade and a half, issues of limited location adaptability have plagued many of the projects I have worked on. I was required to deal with these limitations on my projects before I discovered that this functionality had been developed. Therefore, for many years I had a consistent problem but no software solution. However, I can now say confidently that these issues can be addressed with the software from one vendor.

Let's take the three superplant functionalities as an example. In environments where there are dependencies between production activities—such as when a finished good in one factory is fed by semi-finished goods or components (or the components are in a third factory feeding the semi-finished goods plant, which I have in fact seen at several companies)—then the production across the various plants ends up being poorly integrated. For example, the widely-used SAP PP/DS application, which I have implemented, has no way to resolve conflicts between plants that feed each other. Like almost all production planning and scheduling applications, the application cannot even "see" this relationship because its design is such that each plant is seen as *independent* during the planning run.[1] While the supply planning system can see the overall supply network, the vast majority of supply planning software can do nothing with respect to multi-plant planning because it cannot create routings that span multiple factories.

Essentially, what has occurred in this scenario is that there is *one long production line*, which is divided between different plants. These plants could be located close to one another, or as in the example of one of my clients, on different sides of the globe. In an environment of increasingly globalized manufacturing, a very long production line that spans the globe is more common than ever.

---

[1] SAP SNP (Supply Network Planning) can see the overall network and works with SAP PP/DS, but as will be explained in Chapter 4: "Single Versus Multi-pass Planning," sequential supply and production planning runs will not provide multi-plant planning functionality.

There are tricks that can be performed in standard applications that lack multi-plant planning functionality (I chronicle them in my book *Multi Method Supply Planning in SAP APO*) to make the planning systems with a traditional design do the best possible job of integrating these processes. Fundamentally, most supply and production planning software in the market was never designed to meet these requirements; most are designed with a purely sequential approach between supply planning and production planning. The supply plan is created first, as is shown in the graphic below:

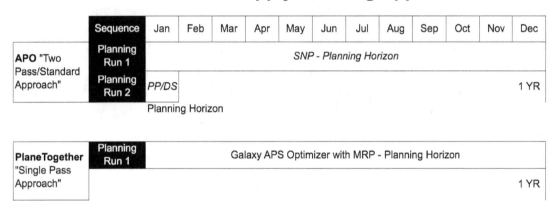

Production and Supply Planning Approach

	Sequence	Jan	Feb	Mar	Apr	May	Jun	Jul	Aug	Sep	Oct	Nov	Dec
APO "Two Pass/Standard Approach"	Planning Run 1						SNP - Planning Horizon						
	Planning Run 2	PP/DS											1 YR
		Planning Horizon											

	Planning Run 1												
PlaneTogether "Single Pass Approach"	Planning Run 1					Galaxy APS Optimizer with MRP - Planning Horizon							
													1 YR

Few vendors have designed their products to meet a requirement like multi-plant planning, or to completely integrate production and supply planning runs.

This book attempts to clearly define specific advanced adaptive functionality: multi-plant planning, subcontracting and multi-source planning. Multi-plant planning is at the intersection of production and supply planning. It is software functionality that has been known as "multi-plant" or less frequently as "multi-site" functionality. However, the term "multi-plant," while well-known by some of those knowledgeable in production planning software, has never really struck a chord outside of that relatively small group of specialists. It has also not been adopted by many software vendors. In my informal queries on projects, I have found that the vast majority of those who either manage production planning implementations or work with production planning software do not know what the term means. Even for those who have read up on the topic, the term multi-plant has been diluted by a number of software vendors. The most prominent diluter of the term has been SAP, which has adopted the term "multi-plant" to describe the functionality that is interplant supply planning. It's a strange adoption/corruption of the term because all supply planning software can source between supply network locations. Multi-plant planning, on the other hand, means that the software can look across plants, compare alternatives, and choose among alternative routings in order to make the best decision under a particular scenario of where to produce. The stock transfers that are the outcome of this process are a result of the ability to perform a comparison across two or more alternative routings. In contrast, single-plant planning (which is by far the most common software design used by production planning software) cannot do this, and can only choose among alternative routings *within* each factory.

Planning Release

Each application/process passes a plan to the next application/process.

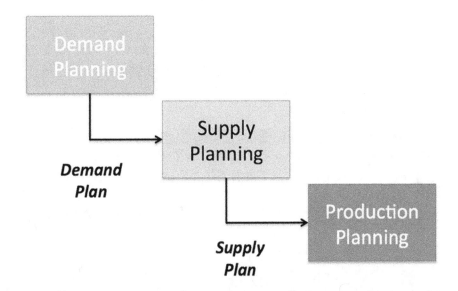

The sequential and "supply planning first" approach means that supply planning is performed prior to production planning and controls the vast majority of the planning horizon. In some applications, supply planning can take into account production resources; however it cannot see the same level of detail as the production planning applications. Therefore the supply planning application creates an oversimplified—or what I call an initial—production plan.

SAP is by far the most popular website returned when one types *"multi-plant planning"* into Google. In fact, when I tried this search, SAP's website(s) were the top six search results as shown in the screenshot on the following page:

Multi-Plant (Site) Planning

help.sap.com/saphelp_40b/helpdata/en/f4/.../content.htm

Multi-Plant (Site) Planning. Using **multi-plant planning**, you can carry out material requirements planning for various plants centrally. This planning procedure ...

Multi-Plant/Site Planning (SAP Library - Material Requirements ...

help.sap.com/saphelp_46c/helpdata/en/f4/.../content.htm

Using **multi-plant planning**, you can carry out material requirements planning for various plants centrally. This planning procedure facilitates the production of a ...

Shaun Snapp shared this in Gmail – Limited

Multi-Plant Planning with Stock Transfer

help.sap.com/saphelp_40b/helpdata/en/7d/.../content.htm

Multi-Plant Planning with Stock Transfer. Within the stock transfer procedure, goods are produced and delivered within a company. The plant which is to receive ...

Shaun Snapp shared this in Gmail – Limited

Multi-Plant Planning (SAP Library - Planning Table (PP-REM))

help.sap.com/saphelp_erp2004/helpdata/en/70/.../content.htm

Multi-Plant Planning Locate the document in its SAP Library structure. Purpose. You use this process if you want to plan and then produce a finished product in ...

Multi-Plant (Site) Total Planning

help.sap.com/saphelp_45b/helpdata/en/d7/.../content.htm

Multi-Plant (Site) Total **Planning**. Use. To avoid having to **plan** each plant individually, you can **plan** as many plants as you want in one total **planning** run.

Shaun Snapp shared this in Gmail – Limited

Below are links to SAP Help that describe multi-plant functionality, but which have little to do with actual multi-plant functionality.

http://help.sap.com/saphelp_45b/helpdata/en/d7/5c9366f47811d1a6ba0000
e83235d4/content.htm

http://help.sap.com/saphelp_40b/helpdata/en/7d/c276fc454011d182b40000
e829fbfe/content.htm

http://help.sap.com/saphelp_46c/helpdata/en/f4/7d2d4d44af11d182b40000
e829fbfe/content.htm

Books and Other Publications on Superplant Functionalities

"Superplant" is a term which I created, so one would not expect other books to use this terminology. Superplant is based upon three functionalities: multi-plant planning (primarily covered in academic publications), multi-sourcing (primarily covered in some academic publications and in vendor documentation), and subcontracting (covered—from a software perspective—primarily in vendor documentation).

As with all my books, I performed a comprehensive literature review before I began writing. One of my favorite quotations about research is from the highly-respected RAND Corporation, a "think tank" based in sunny Santa Monica, CA. They are located not far from where I grew up. On my lost surfing weekends during high school, I used to walk right by their offices with my friend—at that time having no idea of the institution's historical significance. RAND's *Standards for High Quality Research and Analysis* publication makes the following statement about how its research references other work.

> *"A high-quality study cannot be done in intellectual isolation: It*
> *necessarily builds on and contributes to a body of research and*
> *analysis. The relationships between a given study and its predecessors*
> *should be rich and explicit. The study team's understanding of*
> *past research should be evident in many aspects of its work, from*
> *the way in which the problem is formulated and approached to the*
> *discussion of the findings and their implications. The team should*

take particular care to explain the ways in which its study agrees,
disagrees, or otherwise differs importantly from previous studies.
Failure to demonstrate an understanding of previous research lowers
the perceived quality of a study, despite any other good characteristics
it may possess."

The few books that cover these areas of functionality spend only a few pages on them. And providing more coverage was something I very much looked forward to doing with this book; these topics are complex and require the significant coverage I was able to allocate to them.

I have first-hand experience with some of the vendors that have published the most material on both subcontracting and multi-sourcing—and I disagree with the design approaches most software vendors have taken. As I've seen their products fail quite badly on projects, I will use their poor functionality only as examples of what not to do. It just so happens that the examples of poorly-designed functionality will include one of the best-known software vendors in enterprise software.

The story of multi-plant planning functionality is a bit more straightforward, as I am not aware of more than one vendor that can implement multi-plant functionality—and interestingly, their application was designed to perform multi-plant planning right from the beginning.

The Origin of the Term "Superplant"

The origin of the term "superplant" is easy to explain, because I coined the term. The term actually came to me as I was touring factories outside of the US and discussing the requirements of a previous client who needed their software to manage interdependent factories. During the factory tour I remarked to one of the SAP APO resources that:

"It's as if the company's factories are simply individual workstations
in a global supply network......like one giant superplant."

A few other members of the project adopted the term, and I went about developing articles on SCM Focus explaining the concept.

http://www.scmfocus.com/sapplanning/2012/07/26/the-superplant-concept/

http://www.scmfocus.com/productionplanningandscheduling/2013/04/22/multi-plant-superplant-planning-definition/

The Ability to Plan a Single Virtual Plant (for at least some products)

Multi-plant planning is a radical concept because its functionality disrupts the traditional division of responsibilities between the supply planning system and the production planning system. For software to meet the criteria of being multi-plant capable, it must be able to do the following:

1. It must have the ability to create alternate routings that span locations. Multi-plant planning has the ability to treat all resources and all plants as if they are part of one enormous "superplant." This functionality is activated selectively only when it meets the business requirement. Interestingly, with the vendor showcased in this book, resources must actively be assigned to a plant; they are not naturally created within a plant, which is the traditional design approach.

2. It must have intelligence that can allow it to make decisions among alternative routings. Hypothetically it's possible to have some other method, but it is difficult to see this method being anything but an optimizer.

My introduction to multi-plant planning was on a project for a multinational company which produced various components that went into their finished good in different factories separated by more than 4,500 miles. At this time I had never heard of the term multi-plant planning, and was not aware of the previous academic research in this area. My first objective on this project was to simply do what I could in SAP APO (SAP's advanced planning application suite) to meet this multi-plant planning requirement. APO was the tool my client had selected, so that is what I needed to get to work properly; we were going to use a combination of SNP and PP/DS to meet the requirement. Some of the information regarding how this was done was published in the book *Multi Method Supply Planning in SAP APO*. (This book actually addresses several topics. One topic is how to use multiple methods effectively to plan a supply network, but another component was designed to specifically address the needs of a superplant.) We had to do quite a bit of work to make it function, but I was never actually satisfied that the approach we selected—which was based upon adjustments to the supply planning system which customized how product locations were treated by the system—was the best approach for meeting the requirement. (However, I do believe it is the best possible option for those who have already committed to SAP APO.)

After completing this project I contemplated whether any other software vendors had the actual functionality to address the multi-plant planning requirement. Luckily, due to my work either writing articles with best-of-breed vendors or working with them in other capacities, I own a number of their manuals. I decided to read through PlanetTogether's manuals for their application Galaxy APS and checked more on the topic, and I realized that the requirements that I had just made a great deal of effort to test in software with an older software design were covered by **standard functionality** in PlanetTogther. Interestingly however, few of the companies that used PlanetTogether actually leveraged this functionality.

The topic of the superplant was supposed to be part of my book *Constrained Supply and Production Planning in SAP APO*. However after forwarding sections of that book that dealt with PlanetTogether to that vendor, the concept seemed to dovetail with their efforts to raise awareness among companies about performing planning in this way. Therefore, I decided to remove the chapter on the superplant from *Constrained Supply and Production Planning in SAP APO* and to turn the topic into its own book. Doing so allowed me to provide much more detail on topics that are of course the chapters of this book. It also allowed for a distinct book in an area where there were none. Furthermore, up until this book, all of my books have been focused primarily on how to set up software to meet business requirements. This is the first book that can be read by executives as well as those interested in the technical aspects of modeling and planning a superplant.

Future of the Term "Superplant"

If the term "superplant" becomes popular, it will no doubt be diluted because software marketers never rest and the competition for software sales is so great. Clear and accurate terminology is to the advantage of those vendors who already have the functionality in question (and the consulting companies aligned with these vendors), but it's disadvantageous to all the other vendors. Those vendors that do not have the functionality (and the functionality is not easy to add to an existing system) will of course look to confuse their potential clients on the topic. Confusing the terminology is one strategy, but not the only strategy that can be employed when functionality is missing. Another strategy is to say that the functionality is not important. This was a major strategy employed by SAP in the late 1990s when companies were becoming interested in advanced planning. At the time SAP had ERP systems to sell, but no advanced planning software. Therefore they told clients that advanced planning was not very important. A third strategy for a vendor to use to counteract a disadvantage in functionality is to propose that, while the functionality may be beneficial, the time is not right for the functionality or for the company to deploy the functionality because there is so much else to do—the company has *"bigger fish to fry."*

I should note that the latter two strategies can both be one hundred percent accurate. Advanced functionality can be enticing, but of less practical importance in the real world than it appears to be conceptually. Secondly, it is also true that a company may find functionality enticing that it is simply not ready to deploy. I have made both of these arguments on numerous occasions myself. However, at the time it was my honest opinion. The difference with the use of these strategies by software vendors who lack functionality is that the opinion *is developed to minimize a competitive weakness*, and may not be the legitimate opinion of the person proposing it.

At the time of this book's publication, the overwhelming majority of companies that have superplant requirements are meeting them *without software that can perform superplant planning*. Some of these companies are ready for superplant implementations and some of them are not. Furthermore, companies that do have superplant requirements are on a continuum between those companies that require minor superplant functionality and those companies with high superplant requirements.

The Applications Showcased in This Book

This book will describe the superplant concept, which is an overall adaptive supply and production planning business requirement that is met by very few supply chain applications.

Two different applications are explained in this book. One is SAP APO and the second is PlantTogether. SAP APO is a broad advanced planning suite that includes, depending upon how you count, around ten modules (although the bulk of its implementations are in just four modules). The relevant modules for supply planning and production planning are called Supply Network Planning (SNP) and Production Planning and Detailed Scheduling (PP/DS) respectively. As with all of the APO modules, SNP and PP/DS work in conjunction with each other. While there is some overlap (which will be explained), these are traditional sequential and supply planning first applications. Their design goes back to the mid 1990s, and both products are patterned off of products made by i2 Technologies, one of my former employers. SAP APO is used as an example of the standard non-multi-plant production planning system and for the purposes of explaining how the vast majority of supply and production planning systems work. SAP APO's subcontracting functionality is very difficult to use, as is its multi-sourcing functionality. SAP APO is not alone in this regard, as subcontracting planning and multi-sourcing are areas that are commonly problematic on advanced planning projects. This is due to the fact that few advanced planning applications have been designed from the beginning to treat locations flexibly as if they are *either* internal *or* external to the enterprise which implements the software. Therefore these planning systems run into problems when the locations are pseudo internal or pseudo external depending upon how one wants to look at it. Unfortunately for many implementing companies, this is increasingly where requirements are going, meaning that many supply and production planning systems are out of date with these requirements. This mirrors the problems that have plagued software vendors with respect to collaboration requirements. As with the treatment of locations, the treatment of collaboration in the design of most supply chain planning applications has been poorly conceived.[2] Thus, the problems SAP APO has performing both these functions is a useful explanation as to the problems that many other software vendors share.

A natural question might be: If SAP APO lacks multi-plant functionality, why is it showcased in this book? The reason is simply that without understanding

---

[2] This topic is related to collaboration design in software, which is barely discussed. It is covered in the following article: http://www.scmfocus.com/supplychaincollaboration/2013/07/why-must-specialized-supply-chain-collaboration-applications-exist/

the traditional supply planning and production planning design, it is difficult to understand how a superplant-compliant application differs. I anticipate that very few people who read this book currently work with an adaptive application like PlanetTogether's Galaxy APS. Therefore, SAP APO provides me with the *perfect counterweight* to Galaxy APS.

I have worked in advanced planning since 1997. I have implemented SAP APO since 2003 and have written five books on APO. I have worked with PlanetTogether since roughly 2010, and have written this book as well as *Process Industry Planning Software: Manufacturing Processes and Software,* which is another book that covers PlanetTogether but focuses on an area greatly underserved by production planning and scheduling software. Therefore, I have a long-term and practical exposure to both applications.

PlantTogether As *the* Superplant Application

The second application to be showcased in this book is called PlanetTogether (PT). I selected PT based upon the fact that it is one of the few production planning applications that can presently perform multi-plant planning. In addition, PT has subcontracting/contract manufacturing functionality that is both usable and straightforward to configure.[3] I became familiar with PT several years ago—both the company and the software—after writing a number of articles on the software. These articles are at the following link:

http://www.scmfocus.com/productionplanningandscheduling/

However, even though PT is capable of supporting the superplant requirement, it should be understood that as of the publication of this book, multi-plant planning is new and usually *PT is not implemented using the multi-plant functionality. It is sometimes implemented with either its subcontracting or multi-source planning functionality activated*. SCM Focus views this as a big missed opportunity for many companies. One of the major reasons for companies not implementing this functionality is simply the customers' lack of knowledge of these functionalities, something I hope this book is able to address.

---

[3] Galaxy APS does not, however, perform multi-source planning.

A company may implement Galaxy APS because it is attracted to Galaxy APS's adjustable duration-based optimizer, or how Galaxy APS handles process industry requirements, or because of its flexibility with regards to master data maintenance, or because of its user interface. As I will show further on in this book, many companies that have implemented Galaxy APS do not have superplant requirements and so they do not investigate or have a particular interest in whether or not Galaxy APS contains this functionality. Therefore, I do not want to leave the reader with the impression that superplant planning is the primary function of Galaxy APS, or the main reason it is chosen by companies for implementation, because that is not the case.

Being able to compare and contrast two systems that do supply and production planning and yet work in such different ways is, I think, a main contribution of this book. If I had stopped at showcasing only how multi-plant planning works, I believe the audience for the book would have been much smaller.

Galaxy APS User Interface Introduction
On the next page is the main user interface to Galaxy APS.

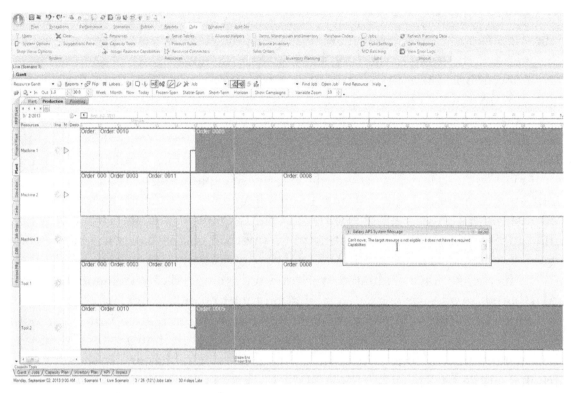

"The system will only schedule work on eligible Resources, and Drag-and-Drop is only allowed to eligible Resources (or a rejection warning is shown),"[4]

Publications on Multi-plant Planning and the History of Method Development

It was quite interesting to review the literature on multi-plant planning necessary to write this book. Clearly, multi-plant planning is an emerging area because most of the substantial documentation on the topic is in the form of academic papers. I have studied the history of a wide variety of supply chain planning methods for other books. From this research I have found that, in most cases, the academic documentation precedes the software development. (Some might think

---

[4] PlanetTogether APS Product Training.

that this would always be the case, but in fact it is not so—as I will explain in just a few sentences.) Furthermore, it is typically the case that the more complex the method, the greater the time lag between the appearance of the method in academic publications and the implementation of the method in software. For instance, inventory optimization and multi-echelon planning—one of the major methods of supply planning—first appeared in an academic paper in 1958, but was not implemented in commercially available software until the late 1990s. There are several reasons for this time lag. Often academics develop mathematics for a method that is so far beyond the hardware capabilities of the time, that the wait for a commercial application will be a long one.[5] Academics generally strive for pushing the envelope. For instance, in the examples of inventory optimization and multi-echelon planning, testing of the method on the hardware of the time could only be performed for a limited supply network and for a single product. Of all the methods I have evaluated over time, only material requirements planning (MRP) and, to a lesser degree, distribution requirements planning (DRP) break this rule of the transition from academic research to commercialization. In both cases the method was developed first in software and in industry, and then was documented in academic or academic-type papers. This history is shown in the graphic on the next page:

---

[5] Multi-echelon planning research was funded by the Air Force through the RAND Corporation. One of the interesting questions is why the Air Force was interested in funding research that was so far beyond the hardware capabilities of the time that it would be decades before it could be used. This research funding by the Air Force reinforces points made by Noam Chomsky, that a misunderstood feature of US military spending is that it is a clandestine way of directing research money to areas of technology that are in fact part of an industrial policy. Curiously, the Air Force was willing to fund the development of the mathematics for multi-echelon planning, but not for the development of software to actually run a multi-echelon software program—another peculiarity of how it distributed research funding.

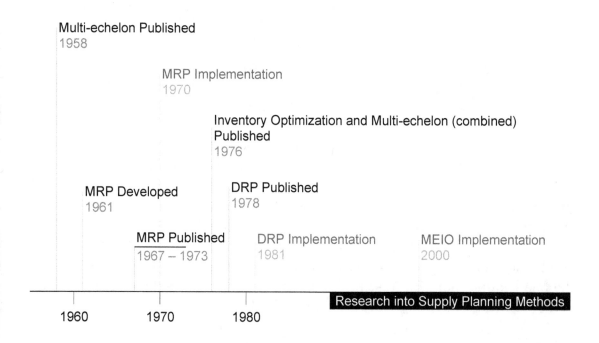

Outside of academic publications and several interesting examples of nonacademic publications that I discuss in this book, the other most prominent publishers on this topic are SAP (which I have already explained describes something that is not multi-plant planning yet uses the term "multi-plant planning") and PT. PT has begun to write about the topic on its website, and they cover multi-plant planning in their training manuals. (This book was written with input from PT.)

Some of the websites that describe multi-plant planning are not discussing "true" multi-plant planning, but some other use of the term. Therefore, even though they use the terminology, they don't actually cover the topic according to the officially accepted use of the term. Naturally, I do not count them as contributors in this space.

Generally, what I came away with from my research was that the term "multi-plant" was used lightly, and the concept behind it only weakly explained and socialized. I faced a similar issue—although to a lesser degree—with my book on Inventory Optimization and Multi-echelon Planning (MEIO). This was a method, functionality, and mathematics for which a number of vendors had already developed products, and for which quite a bit of academic literature had been written. However, before my book and sub-site on the topic at SCM Focus, there was no real comprehensive explanation for business people as to how MEIO worked outside of the software manuals of the individual vendors.

At the time of this book's publication, multi-plant planning is at a far earlier stage of development than anything I have written about in the past. For an author like myself, who enjoys bringing the latest innovations in supply chain software from the best vendors, this is exactly the kind of topic—and the stage of development of the topic—that I look for.

Of all the publications I read on the multi-plant topic, it is no coincidence that the one I found more interesting is one of the important publications on multi-plant planning: *An Analytical Framework for Multi-Site Supply Chain Planning Problems*. It has some interesting quotations and frameworks. Several of the ones that I consider the most impactful I have included and commented upon below:

> *"In the past years the multi-site production planning problems have attracted many researchers' attention, but most of the researches put emphasis on the methodology to solve the multi-site planning and scheduling problem. Few of those researches are to analyze the essence and definition of the multi-site production planning problem. The analytical framework of the multi-site production planning problem is proposed in this paper."*

This is an important point because understanding the "analytical framework" essentially means understanding how multi-site or multi-plant planning actually works. The paper also discusses the "structural framework," which I have listed below, along with a description of how each aspect of the structural framework is covered in this book:

- *Multi-site Conceptual Model:* How multi-site planning works from a design perspective is very much a focus of this book.

- *Product Structure (Bill of Materials), Resources and Routings:* How each one of these master data objects must operate and be set up in order to support multi-plant planning. The bill of materials (BOM) is covered to the degree that one overall finished goods BOM must be broken into modular BOMs in order to produce components of the finished good in separate factories. Resources and routings are covered quite extensively.
- *Production Methods:* The planning method used changes how the production and supply planning model must be set up in order to achieve the desired outcome. This book will discuss cost optimization, duration optimization, and duration optimization with a heuristic-based algorithm.
- *Manufacturing Capability and Characteristics:* This book addresses capabilities that are assigned to resources. Understanding how one capability can be assigned to multiple resources that are in different plants is critical to understanding how superplant functionality works.
- *Production Planning Constraints:* This book will assume that constraint-based planning is used in all cases. The most important concept to understand with respect to the use of superplant planning along with constraint-based planning is that a single bottleneck/drum resource may constrain the entire production process regardless of how many plants the routing is strung through.[6] This is a focus of coverage in this book.
- *Key Performance Indicators (KPIs)*:* Optimizing with KPIs is a big part of what makes PlanetTogether work so effectively. Not only will KPIs improve when multi-plant planning functionality is properly deployed, but the Galaxy APS optimizer can be adjusted through the use of KPI adjustment rules. This will be explained in detail in several places in this book.

Different Multi-plant Planning Scenarios

An Analytical Framework for Multi-Site Supply Chain Planning Problems goes on to say the following about how the definitions of multi-plant are used generally.

> *"In the literatures, many researchers had different definitions for the term: 'multi-site' or 'multi-plant.' The multi-site production planning*

---

[6] To understand constraint-based planning in a way that effectively interoperates with this book, but with a focus on where there is a separate supply and production planning system, please see the book *Constrained Supply and Production Planning in SAP APO.*

problem is mainly the production allocation decisions among multiple plants. (Thierry et al.) ...manufacturing process of products may require the usage of many resources located in different production units. Furthermore, some alternative manufacturing routings may exist."

The paper points out that there are multiple multi-plant models, and so it focuses only on the major multi-site models. The paper then goes through the various major models, which are essentially different scenarios. One scenario may have stock transferred between two factories. Another may have stock coming in from one location, or from multiple locations to two or more factories. However, for our purposes, if the requirement were to synchronize production and supply planning across multiple locations, on components that eventually lead to a combined product—most often a finished good, but also an assembly or subassembly—then this would meet the definition of multi-plant planning. When this is the case, the company is in some shape or form planning a "superplant."

The Opportunity of Multi-plant Planning

There is little doubt of the many companies with multi-plant planning requirements. I have experienced the requirements first hand and in depth at one company, but have found these requirements at other companies as well—it's simply that most are unaware what these requirements are actually called and that there is academic work that describes these requirements. They are also unaware that they are losing efficiency by not being able to account for these requirements. Multi-plant planning is an important stage in the evolution of planning software that is related to subcontract and contract manufacturing planning. All of these planning requirements mean the planning system is indifferent to whether the manufacturing location is owned or not owned by the implementing enterprise. Companies have little in the way of information on multi-plant planning, which should not be surprising in the least. In fact, I know from my consulting experience that many companies have multi-plant requirements but are simply not leveraging the software currently available to manage these requirements. In fact, at most companies the internal discussion about doing so has not even begun. Common reasons as to why this is the case are listed below:

1. Many decision makers in companies with multi-plant planning requirements do not know that the functionality to specifically address these requirements exists.

2. Many companies do not include vendors with multi-plant planning functionality in their software selection initiatives.

3. No ERP vendor makes external planning software that performs multi-plant planning. The company would have to be willing to choose a smaller vendor rather than simply purchasing the ERP vendor's external planning system. This is of course a limiting factor, because companies tend to purchase as much software as they can from one vendor, which incidentally is why the enterprise software sector is so monopolistic in nature and why so many companies have such a poor fit between their business requirements and the applications they have purchased.

Furthermore, as companies have moved to more specialization in their factories (co-locating specific manufacturing in global locations), intercompany transfers have become increasingly common. Concentrating similar types of production in factories globally has been occurring for some time. Things like subcontracting, which at first glance would seem to reduce the necessity for multi-location planning, in fact increase the necessity for multi-location planning. Even in instances where third parties are involved—such as with subcontracting—the primary company or OEM *often wants to plan the activities,* even if they do not perform the actual execution. In fact, we now have the common scenario where planning factories—or at least partially planning factories that *are not owned by the company performing the planning—are a common requirement.*

As most enterprise software is not designed to accommodate these requirements, a great deal of effort is needed on the part of companies to both implement and maintain the software. The retort from many vendors might be, *"but we offer supplier collaboration and subcontracting"*—which is true. However, it is also true that these tend to be tricky implementations, and in some cases, such as with supplier collaboration, there are in fact few success stories.

http://www.scmfocus.com/supplychaincollaboration/2010/06/where-are-the-supply-chain-collaboration-success-stories/

Therefore, the business requirements and opportunities for multi-plant planning will only continue to grow. In fact, this is a constant trend in both supply and production planning. Supply chains are simply becoming more global with more interactions of all types between plants. Perhaps more importantly, in the case of locations that are not part of the company's supply network, supply chains are also becoming more *ambiguous as to who is actually performing the planning*. The business needs within these companies have changed more quickly than the software that supports them.

Subcontracting

Along with multi-plant planning, subcontracting has greatly increased as a planning need within companies. Subcontracting is another form of production where there is ambiguity between the external plant and the internal locations. Another related concept is contract manufacturing, where the product is produced completely by the contract manufacturer but planning responsibility is shared. Some supply chain planning applications can plan subcontracting; however, the functionality I will describe in this book can make comparisons between alternative production that is either internal or external to the company. In some circumstances, production may be outsourced; in other cases it may be planned to be produced internally. Oftentimes inflexible planning systems mean that companies are forced to make these types of decisions "strategically." However in the software described in this book, the alternatives can be set up in the model, and the application can switch between internal and outsourced manufacturing as the situation changes. I cover contract manufacturing along with subcontracting because there are a number of similarities and the functionality that is explained in this book can meet either requirement.

Multi-source Planning

Multi-source planning (or multi-sourcing)—the ability to have the system flexibly choose from external sources of supply—has been a consistent requirement at many companies. However, many companies have also had a problem getting multi-source planning to work the way they want it to work, and so it is not implemented as commonly because of issues with functionality robustness. I have spent years dealing with these issues on projects—projects that have required enormous amounts of configuration and testing because essentially the software I was using was not designed to work the way that the company needed it to work.

I am quite enthusiastic about the business benefits that can be derived from a fully superplant-capable application. But from a purely selfish perspective, I would simply like to be able to implement this software, because the constant conversations about the inability of various applications to meet these needs (and the workarounds that they entail) have become a headache to me as much as to my clients. Therefore, a motivation for writing this book is to explain to the widest group possible that it is not necessary to continue to bang our collective heads against the wall and perform various magic tricks (which always seem to impact some other functionality) on projects in order to meet these requirements. All of the adaptive functionalities mentioned can be implemented both effectively and cost effectively with PT's Galaxy APS product.

The Use of Screen Shots in the Book

I consult in some popular and well-known applications, and I've found that companies have often been given the wrong impression of an application's capabilities. As part of my consulting work, I am required to present the results of testing and research about various applications. The research may show that a well-known application is not able to perform some functionality well enough to be used by a company, and point to a lesser-known application where this functionality is easily performed. Because I am routinely in this situation, I am asked to provide evidence of the testing results within applications, and screen shots provide this necessary evidence.

Furthermore, some time ago it became a habit for me to include extensive screen shots in most of my project documentation. A screen shot does not, of course, guarantee that a particular functionality works, but it is the best that can be done in a document format. Everything in this book exists in one application or another, and nothing described in this book is hypothetical.

Identification of Timing Field Definitions

This book is filled with lists. Some of these lists are field definitions. Lists of field definitions will be all *italics*, while lists that are not field definitions will be only *italics* for the term defined, while the definition that follows is in normal text.

How Writing Bias Is Controlled at SCM Focus and SCM Focus Press

Bias is a serious problem in the enterprise software field. Large vendors receive uncritical coverage of their products, and large consulting companies recommend the large vendors that have the resources to hire and pay consultants rather than the vendors with the best software for the client's needs.

At SCM Focus, we have yet to financially benefit from a company's decision to buy an application showcased in print, either in a book or on the SCM Focus website. This may change in the future as SCM Focus grows—but we have been writing with a strong viewpoint for years without coming into any conflicts of interest. SCM Focus has the most stringent rules related to controlling bias and restricting commercial influence of any information provider. These "writing rules" are provided in the link below:

http://www.scmfocus.com/writing-rules/

If other information providers followed these rules, we would be able to learn about software without being required to perform our own research and testing for every topic.

Information about enterprise supply chain planning software can be found on the Internet, but this information is primarily promotional or written at such a high level that none of the important details or limitations of the application are exposed; this is true of books as well. When only one enterprise software application is covered in a book, one will find that the application works perfectly; the application operates as expected and there are no problems during the implementation to bring the application live. This is all quite amazing and quite different from my experience of implementing enterprise software. However, it is very difficult to make a living by providing objective information about enterprise supply chain software, especially as it means being critical at some point. I once remarked to a friend that SCM Focus had very little competition in providing untarnished information on this software category, and he said, "Of course, there is no money in it."

The Approach to the Book

By writing this book, I wanted to help people get exactly the information they need without having to read a lengthy volume. The approach to the book is essentially the same as to my previous books, and in writing this book I followed the same principles.

1. **Be direct and concise.** There is very little theory in this book and the math that I cover is simple. While the mathematics behind the optimization methods for supply and production planning is involved, there are plenty of books, which cover this topic. This book is focused on software, and for most users and implementers of the software the most important thing is to understand conceptually what the software is doing.

2. **Based on project experience.** Nothing in the book is hypothetical; I have worked with it or tested it on an actual project. My project experience has led to my understanding a number of things that are not covered in typical supply planning books. In this book, I pass on this understanding to you.

3. **Saturate the book with graphics.** Roughly two-thirds of a human's sensory input is visual, and books that do not use graphics—especially educational and training books such as this one—can fall short of their purpose. Graphics have also been used consistently and extensively on the SCM Focus website.

Before writing this book, I spent some time reviewing what has already been published on the subject. This book is different from other books in terms of its intended audience and its scope. It is directed toward people that have either worked with ERP or know what it is; I am assuming that the reader has a basic knowledge level in this area.

The SCM Focus Site

As I am also the author of the SCM Focus site, http://www.scmfocus.com, the site and the book share a number of concepts and graphics. Furthermore, this book contains many links to articles on the site, which provide more detail on specific subjects. This book provides an explanation of how supply and production planning software works and aims to continue to be a reference after its initial reading. However, if your interest in supply planning software continues to grow, the SCM Focus site is a good resource to which articles are continually added.

The SCM site dedicated specifically to supply planning is
  http://www.scmfocus.com/ supplyplanning.

The SCM site dedicated specifically to production planning is
  http://www.scmfocus.com/productionplanningandscheduling/

The site dedicated to SAP planning is
  http://www.scmfocus.com/sapplanning

Intended Audience

This book was written for those with an interest in leveraging leading approaches in the supply network for planning improvement. The specific audience would range from executive decision makers to software implementers to supply and production planners. If you have any questions or comments on the book, please e-mail me at shaunsnapp@scmfocus.com.

Abbreviations

A listing of all abbreviations used throughout the book is provided at the end of the book.

Glossary

The following glossary of terms will be helpful before reading Chapter 3: "Multi-plant Planning" and some have been taken from PT literature. It can be best to breeze through these terms when first reading this chapter, but then return to them when the terms are first used.

1. *Plants*: Correspond to separate physical factories (I use the terms "plant" and "factory" interchangeably).

2. *Resources*: Usually correspond to machines, people, and tools or subcontractors. They are used temporarily to perform an operation. Materials are not Resources since they are "consumed" rather than used temporarily. Each Operation in a Job has one or more Resource Requirements, each of which specifies the need for another additional Resource in the production process. Each Resource belongs to exactly one Department. A Resource is a capital asset that is available and used to perform work. In this broad definition of a Resource, Galaxy APS considers any machine, tool, die, mold, or even skilled laborer as a resource. The general rule is that if it is a production asset that requires scheduling, it will be defined as a Resource inside Galaxy APS.

3. *Routing*: A combination of resources in sequence. (As will be shown, a major component to superplant functionality is how routings can be configured.)

4. *Alternate Path*: An Alternate Path specifies the precedence relationships between Operations thus indicating the path that is followed through the shop to produce a product. Each Manufacturing Order specifies one or more Alternate Paths and exactly one Default Path. When the Manufacturing Order is scheduled, the Default Path is used. One Operation can have multiple Successors and multiple Predecessors.

5. *Capabilities*: This is the work that a resource can do. A Capability is a specification of a skill or function that can be performed by a Resource. For example, a Capability might be "cutting" or "inspection." Each Resource has one or more Capabilities that indicate the types of work that it can perform. In addition, each Job Operation Resource Requirement has one ore more required Capabilities indicating the Capabilities that a Resource must have to be considered eligible to perform the Operation. If an Operation, for example, requires a CNC machine and a CNC operator, then you would create two Capabilities such as: CNC and CNC-Operator. Then the Operation would have two Resource Requirements—one specifying CNC and one specifying CNC-Operator.

6. *Manufacturing Order*: What is created in order to plan and schedule actual production. A Manufacturing Order is a request to manufacture a specific quantity of a specific Product. Each Job contains one or more Manufacturing Orders. Each Manufacturing Order contains one or more Operations and one or more Alternate Paths that describe the sequence in which Operations should be performed.

7. *Operation*: This is the work performed on a resource. An Operation is a definition of a single processing step in a Manufacturing Order. The Operation specifies values such as the Cycle Time, Set-up Time, expected Yield, Resource Requirements, Material Requirements, etc. Each Manufacturing Order contains one or more Operations. Operations can be "connected" to establish Predecessor/Successor relationships using the Alternate Path of the Manufacturing Order.[7]

8. *Drum*: Used in the in the Drum-Buffer-Rope metaphor of the Theory of Constraints where the drum is the constraint, the buffer is the material release duration and the rope is when the material is released.[8]

Corrections

Corrections and updates, as well as reader comments, can be viewed in the comment section of this book's web page. If you have comments or questions, please add them to the following link:

http://www.scmfocus.com/scmfocuspress/production-books/the-superplant-concept/

---

[7] From the PlanetTogether Glossary.

[8] Galaxy APS can use Drum-Buffer-Rope scheduling if configured to do so.

CHAPTER 2

Understanding a Superplant Conceptually

There are several important differences between a standard production design and a superplant. A good way to begin this chapter is by describing these differences.

1. In a superplant, the bill of material (BOM) is distributed in multiple locations.

2. In a superplant, the production can be placed in multiple locations.

3. Synchronizing the continuous material flow among factories is critical to maintaining production efficiencies. The triggering of stock transports may change depending upon plant proximity. In plants that are within close proximity, the supply planning system may not need to be involved.

   http://www.scmfocus.com/sapplanning/2012/07/24/
   synchronizing-integrated-factories-with-stock-transfers/

4. In a macro sense, each factory can be thought of as a work center, and the flow through the supply network—for internal locations at least—can be considered a routing.

5. An important consideration with the superplant design can be, depending upon requirements, the prioritization of internal over external demand. This means using both constraint-based planning in addition to being able to automatically prioritize the by-demand type. This is addressed in detail in Appendix C: "Prioritizing Internal Demand for Subcomponents over External Demand."

6. A superplant not only has the ability to evaluate whether to place demand onto internal resources, but also to leverage subcontract and contract manufacturing partners in a similarly flexible manner based upon circumstances. A superplant can perform subcontracting, but may also be able to produce the same item internally, meaning that internal production is compared against subcontracting production.

7. A superplant can constrain a routing, regardless of how many plants it passes through based upon a bottleneck/drum resource that resides in any one of the plants.

8. A superplant can nimbly evaluate the best location to source a product automatically and as part of the standard planning run. A superplant can flexibly alter its source of supply based upon the setup of as many sources of supply as actually exist, and allowing the software to make the determination between alternatives.

9. The best way of thinking about the output of superplant planning is that the output depends upon the particular circumstance. In some circumstances one alternative is selected and in another circumstance a different alternative is selected. These circumstances change all along the planning horizon meaning that different alternatives are the best to select at that point in time.

Now that we have covered how a superplant differs from the standard approach to supply and production planning, we will jump into how each of the functionalities that make up a superplant (multi-plant planning, subcontract planning and multi-source planning) actually work in detail.

CHAPTER 3

Multi-plant Planning

As discussed, multi-plant planning, sometimes called multi-site planning, is the ability to model and make decisions to schedule production between alternate internal production locations that can produce the same product. In some cases these locations may have identical capabilities that would produce a finished good; however, in other cases it may be a comparative process that would produce a component or subcomponent that becomes *an input to another production process at a downstream factory.*

Who Requires Multi-plant Planning Functionality?

By definition, companies that have components and subcomponents of final finished goods that are moved between factories have a multi-plant planning requirement. This requirement exists in all manufacturing environments (discrete, repetitive, process batch, process continuous). I once consulted for a repetitive manufacturer that produced wires in some factories, and then sent these wires halfway around the world to become input products to finished goods electrical products.[9] Transporting intermediate product between factories is also common in the petroleum industry. Many of these stock transfers of intermediate product are between facilities owned by the same company, as the oil industry is one of the more vertically integrated of all industries.[10]

A company that does not have multi-plant planning requirements when they start out, will have these requirements as soon as they choose to consolidate one stage of manufacturing to a single location in order to benefit from economies of scale and economies of specialization in that manufacturing process. Producing more similar items at fewer factories will tend to increase production efficiency. (This would be for a laundry list of reasons including the ability to group similar machines together; improve maintainability of many similar machines in one plant; more trained and interchangeable machine operators with similar skills; the ability to purchase and install larger, fewer and more efficient machines; etc.)

---

[9] Generally speaking, not all components that make up a finished good have an equal level of complexity. It can make quite a lot of sense to manufacture lower-complexity components in lower-cost countries and ship the components to higher-cost countries, which can perform the more complex manufacturing operations. Once such a factory is set up and running, it cannot only send its input components to the factory owned by the same parent company, but can have external customers as well.

[10] Vertically integrated companies tend to be a good fit for multi-plant planning software, although what *is* in fact vertically integrated is an interesting discussion. For instance, Wikipedia lists Apple as a vertically integrated firm because they *"...control the processor, the hardware and software. Hardware itself is not typically manufactured by Apple, but is outsourced to contract manufacturers such as Foxconn or Pegatron who manufacture Apple's branded products to their specification."* Therefore, according to Wikipedia, the main definition of "vertical integration" is whether the company controls the process, not—as more generally thought to be the case—whether the company actually performs all of the tasks. As Apple does no manufacturing, they could perform multi-plant planning, but the plants would be external plants.

Plant Design Before

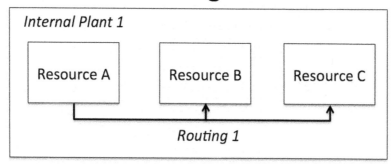

Plant Design After

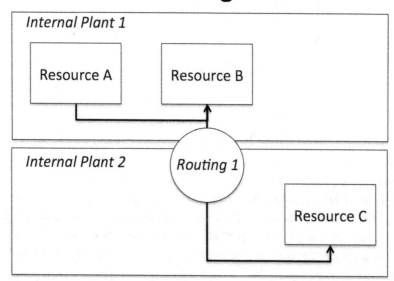

As soon as a routing spans a plant, the company has multi-plant requirements. The more of these relationships a company has, the more it benefits from multi-plant planning functionality.[11]

---

[11] For the purposes of keeping the explanation easy to follow and not combining too many concepts at once, I have left out the discussion of how the stock transfers that are created by the alternate routings between the factories are managed by the solution (the solution being both the external planning system and the ERP system). This is, however, covered in detail in Chapter 8: "The Superplant and the Integration Between ERP and the External Planning System." I consolidated all stock transfer information here so as to cover stock transfers for all three superplant functionalities in one place.

Multi-plant Planning: One Company

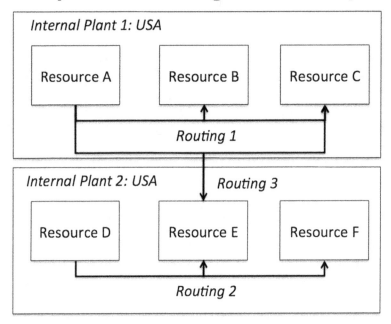

By implementing multi-plant planning software, the company may be able to supply multiple plants. It is able to produce more output with fewer resources because it can receive a higher production utilization from its equipment. Decades of nonmathematical proposals by inventory/lean proponents has blinded companies to the fact that the main emphasis of production is machine efficiency. This can be easily shown using calculations with actual assumptions from factories—these are calculations that lean proponents refuse to perform, relying instead on general statements. Companies that do this receive better production efficiency at the cost of higher planning and coordination costs, as well as higher transportation costs.

There is a tendency to accept current factory configuration as stable. However, throughout the history of manufacturing, factories have gone through many configuration changes. A perfect example of this is during the First Industrial

Revolution, when fossil fuels were first used as a power source over water, wind or even animal power. Also, in the Second Industrial Revolution, steam turbines were moved out of factories and central power plants delivered power to factories over power lines. The power was converted to mechanical movement with electrical motors in the factories. The factories changed enormously before and after both industrial revolutions, not only in their layout but also in their size, specialization, and in almost every other dimension as covered in detail in the article below.

http://www.scmfocus.com/scmhistory/2013/08/the-electrification-of-production-plants/

Factories are simply the result of the technology, material availability, labor skills and costs, and of the industrial engineering and planning principles of the day. They are subject to change, and indeed will change. Companies that have software capable of multi-plant planning are in a better position to leverage the production efficiencies of manufacturing consolidation. Therefore, software capable of multi-plant planning may be implemented by companies that already have multi-plant configurations, or software capable of multi-plant planning may *enable a company to move to a multi-plant configuration only because they have no effective way of properly planning and controlling the factory in an alternate configuration*. Loosely translated, when a company has multi-plant planning functionality, not only can it better manage the multi-plant requirements that it currently has, but can actually adjust its factory configuration to better leverage this newfound planning intelligence.

One example of a multi-plant planning scenario is when a routing (which declares the pathway over the individual operations that the manufacturing/production order must pass) must always span factories in order to arrive at a finished product. However, another example that would drive a multi-plant planning requirement is when identical processing resources existing in multiple internal factories can be leveraged under particular circumstances.

Duplicate Factory

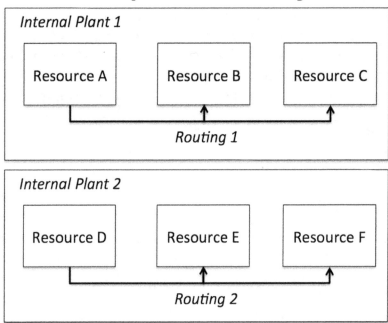

In this scenario there are two duplicate factories that make the same output product. However, when the production rate is particularly high, in one location or the other, the operation that is performed on resource B and E can be "shared." This means making another routing.

Duplicate Factory Extra Routing

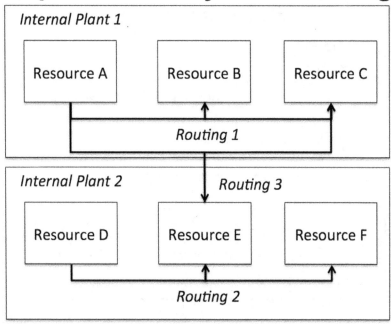

In multi-plant planning, all that is required is to create a new Routing, which allows Plant 1 to call on the capacity of the Resource D in Plant 2. Another Routing could be added which would allow Plant 2 to call on the capacity of Resource E. This may be more efficient than simply satisfying the demand of customers from Plant 2 because considerably more weight is added to the finished product in the final manufacturing step, which is accomplished with Resource C and F.

Duplicate Factory Extra Routing: Subcontracting

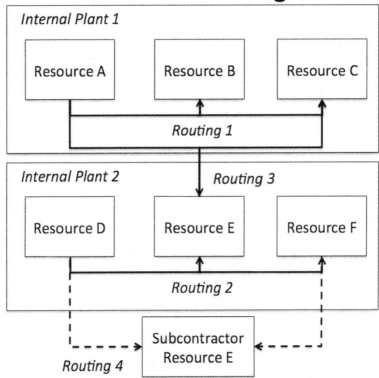

We will get into this topic in Chapter 6: "Subcontracting Planning and Execution." However, with multi-plant planning functionality, we could add another alternate Routing that can reach out to a subcontractor for processing. This would allow Plant 2 to call on the subcontractor for manufacturing capacity. As soon as a company has multi-plant planning functionality, all types of alternative arrangements open up that previously were managed manually.

Understanding the Software Design of Standard Supply and Production Planning Systems

In traditional—and what are by far the most common—supply planning and production planning applications today, if a company were to have one hundred resources spread over ten factories, any resource assigned to one factory is **unavailable to be combined with other resources in other factories**. Under this approach, the supply planning system controls all logic and flow between the locations in the supply network. The production planning activity and capacity within a plant is limited to just that plant.

This software design was clearly colored by the assumption that production planning occurs in a purely local manner. While we have more advanced methods than MRP at our disposal in production planning (including heuristics, allocation algorithms and optimization), it is interesting that the same location limitations of MRP are used even though these more sophisticated methods have location flexibility. For instance, in SNP, as with many supply planning applications, if location decomposition is used with the optimizer, it will break the overall problem into sub-problems with each sub-problem being the location supply network for just the product which is processed. (For more detail on the topic of decomposition see the following article.)

http://www.scmfocus.com/sapplanning/2011/10/12/snp-optimizer-sub-problem-division-and-decomposition/[12]

Each one of these locations can be constrained by all of the production resources in all of the plants—however the production, which is planned, is planned independently in each plant. The supply planning system can see the overall network, but production under this design is planned parochially. Production does not *interact* with the supply plan in a mutually adjusting way; instead, the supply plan decides which plant is to produce what, sends the feasible (if constrained) initial production plan to the production planning system, and then simply leaves the details—such as the detailed scheduling—to be worked out by the production planning system. The two systems are kept in synch when, for instance, adjustments to the production schedule are made in the production planning system, but that is the extent of the interaction.

Once this rather inflexible design approach was adopted, all vendors, roughly speaking, adopted the same assumption, even though this assumption is often at odds with reality. An executive who has been shown the same workflow by multiple

---

[12] For a very detailed explanation of decomposition, see the SCM Focus Press book *Supply Planning with MRP, DRP and APS Software.*

software vendors may naturally question what is wrong with it if so many vendors follow the approach. The reason this design is at odds with reality is because factories do not merely accept raw materials and ship finished goods. Instead, many factories receive raw materials and ship out subcomponents. Other factories receive subcomponents and ship out components or subassemblies. Many possible combinations of factories are possible and always have been. Dated designs like this do not allow the full complement of location interactions to be appropriately modeled within the application. This is a complex set of alternatives, which changes depending upon the circumstances at various points in time. There can be very talented production planners, but there is an upward limit to which any human brain can compare and contrast complex sets of alternatives. Loosely translated, it is the perfect problem to hand over to a computer to solve.

In the history[13] of supply chain planning, methods have generally moved from the oversimplified to the more accurate as time has passed and the ability to model more accurately has improved due to advances in both software and hardware. For example, MRP, at one time the leading edge of supply and production planning, uses a number of highly simplified assumptions in order to generate its planning output. Multi-plant planning is yet another example of this continual increase in modeling accuracy. But, it must also be recognized, particularly by implementing companies as they bear the risks, that more advanced approaches have not always led to more implementable software. In fact quite the opposite is more commonly the case, which is why it is not sufficient to simply introduce a more accurate model or a more sophisticated method if doing so causes the likelihood of implementation success to decline appreciably. Unfortunately, too often this has been the history of new introductions of advanced functionality in all of the supply chain planning software categories. It's hard to emphasize this point enough: being sophisticated is not good enough. And just because a software vendor is offering a new technique or just because other companies have jumped on the bandwagon does not mean it's time to implement the software functionality. Many companies can and have purchased the newest, hottest thing, and have

---

[13] Analysis of history is a major part of our approach at SCM Focus. It relates to everything from our total cost of ownership research to our proposal that it is necessary to assign success probabilities prior to approving implementation budgets.

all failed together. And the really big companies such as Coca-Cola or General Motors seem to fail on their IT initiatives with just as much frequency as any other company—so the fact that some brand name has purchased an application or is implementing a particular functionality means very little. However, through software analysis, this book will show how PT has developed a sophisticated system for providing maximum production location interaction, and one which is also quite implementable.

How Multi-plant Planning Differs from the Standard Design

At SCM Focus we cover multiple areas of supply chain planning software, and through this research we often uncover similarities between the different software categories. For instance, in terms of recent developments in supply chain software, multi-plant planning shares many similarities with the supply planning technique of multi-echelon planning, which I discussed in Chapter 1: "Introduction," in that both methods/technologies expand planning decision-making to be location-agnostic.

With multi-plant planning functionality, each resource is assigned to a particular factory. Routings string together a series of resources through a single factory (the standard approach) or through more than one factory. This requires transportation between factories rather than simple stock movements between workstations within a single factory.

1. In some cases there may be only one routing between more than one plant for a particular product. In that case, the application is performing simple multi-plant planning.

2. In other cases the application may have one or multiple alternate routings—so-called Alternate Paths by PT—and the application must decide upon the best routing/path for a particular circumstance.

3. The application may also compare alternate routings, which are completely contained within one factory, meaning the decision is between routings that string together alternative resources within the plant versus routings which string together resources from multiple external plants. This is not multi-plant planning functionality, and any production planning and scheduling application can do this.

Any number of scenarios is possible. However, the core functionality is first the ability to have routings across factories (this is modeling functionality), and second, a way to make decisions among alternatives (in Galaxy APS this is the optimizer combined with adjustment rules that will be covered in detail further on in this chapter).

Multi-plant Planning Graphically

Below, I use a series of related graphics—beginning at the most simple and becoming more complex—to explain how to gradually build from the simplest scenario (two production lines in a single factory) to the most complicated scenario (multiplant planning across three factories).

Multiple Resources Per Area: Not Interchangeable

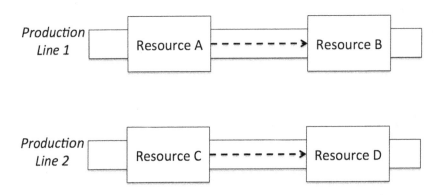

In this first example, we have two production lines in a single factory. Both resources A & C and resources B & D perform the same processing.

Multiple Resources Per Area: Interchangeable:

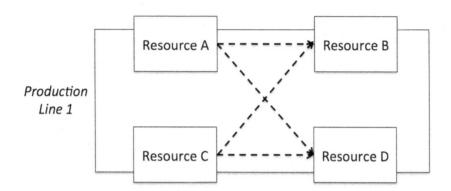

Here, because Resources A & C and Resources B & D perform the same processing, it is possible to aggregate the resources. Resources A & C can feed Resources B &D. This will serve as a baseline for the main point that we want to explain.

Moving to the Next Stage

These first two scenarios shown above have just been a starting point, using the example of resource aggregation within one factory. The following scenarios will not use resource aggregation. Planning level "aggregation" of capacity occurs while at the detailed scheduling level; the work is allocated to specific resources in particular plants. The intention of this clarification is to avoid any misunderstanding on the part of readers that the approach is a bucket-type aggregation of capacity across plants in the scheduling process.

Standard Design

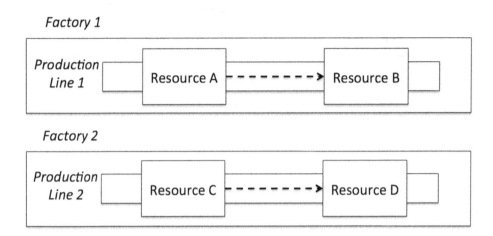

In this scenario, the resources, which are interchangeable, are in different factories. The vast majority of companies plan these locations separately.

Multiple Resources Interchangeable Between Factories

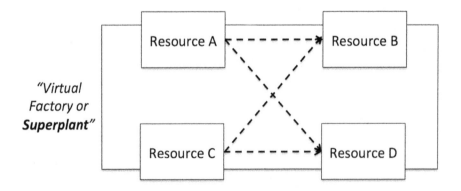

When production planning software has the ability to perform multi-plant planning, the **locations can be planned as a group**.

Multiple Resources Interchangeable Between Factories

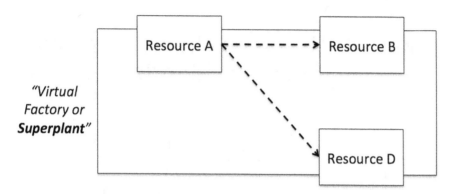

However, with multi-plant functionality, and if there is only one resource in one factory, the resource could service both production lines in both factories.

Multiple Resources Interchangeable Between Factories

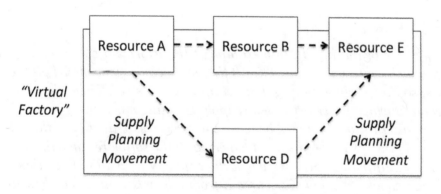

With true multi-plant functionality, the products can move not only to one factory, but also back to the initial factory. These movements are supply planning movements. These supply planning movements are created in response to the requirements that are generated by the selection of the alternate routings within the superplant. The non-multi-plant *planning applications can create stock transfers between locations based upon distribution requirements; however, only multi-plant planning can create stock transfers based upon a selection among alternate routings that are part of an overall production process that runs through multiple factories.*

Multiple Resources Interchangeable Between Factories

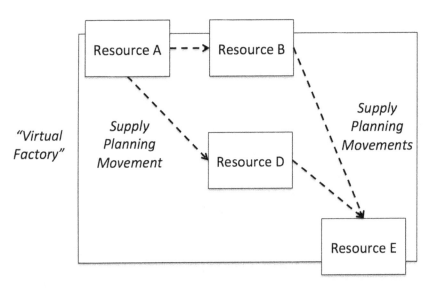

And that is just the beginning. In this example, two factories provide components to a third factory, which contains Resource E. All that is necessary is to model all the possible permutations, so the possible routings are set up in the system. Then, the software will choose the point at which it needs to create supply planning movements between factories when there are alternative resources between the factories. This allows the software to take advantage of the true capabilities of superplant; the software can shift the demand across resources in different factories when the capacity is available. In this example, the software can choose to supply Resource E from either Resource B or Resource D, or both Resource B & D.

Sequential Design (Not Multi Plant)

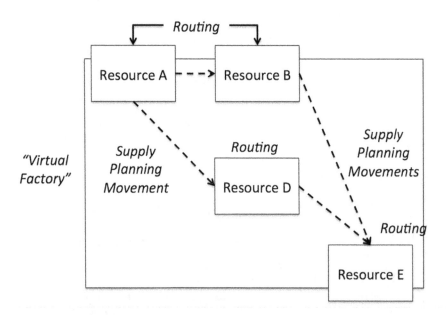

In the non-multi-plant planning design above, notice that each routing is only specific to each location. The supply planning system can pass stock transport requisitions between the locations; however, it cannot plan both supply and production together. Instead the supply plan is created, which then is passed to the production planning system.

Multi Plant Planning

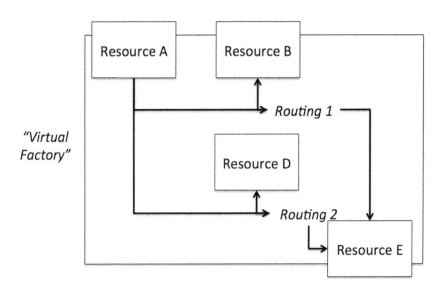

In order for software to be considered multi-plant, routings must be able to span the plants/factories. The multiple routings provide all of the possible permutations that are available within the supply network and the series of plants.

The Criticality of Alternate Paths to the Superplant

The ability to set up and use Alternate Paths within each location is standard production planning functionality, **while Alternate Paths that span multiple locations is multi-plant planning functionality.** PT's training manual has the following to say about Alternate Paths.

> *"Each Manufacturing Order has one or more Alternate Paths that define the sequence of Operations. If there are multiple ways of making a product then an Alternate Path can be created for each method. For example, you may have a product that goes through a lathe operation followed by a grinding operation. In addition, a new CNC machine might be able to create the same product with a single*

operation. This can be accomplished by setting up an alternate BOM/ Routing that gets imported into Galaxy APS as an Alternate Path. One path would have two operations while the other path would have just one operation. (Note that if the only difference between two processes is the production rate on alternate machines then this can be accomplished using Product Rules or a Custom Add-in to customize the rate of the production on each machine.)

Each Manufacturing Order has one Default Path, which is always used during the Optimize process. From the Gantt View, you can manually drag and drop the job onto a different path by using the Alt Key and dragging the Activity Block [Alt + drag]. The alternate resources will be highlighted and you can drag to the alternate path. After dropping the block, if there is more than one Alternate Path that uses that Resource then you will be prompted with a dialog box to choose which Path to use."

However, Galaxy APS can also auto-select Alternate Paths if allowed. The selection is made during optimizations based on the path preferences and the availability of resources used by the various paths.

The Logic for Decision Making in a Multi-plant Environment

Up to this point, we have focused exclusively on the physical modeling aspects of multi-plant planning. However, how the application makes decisions between the many alternatives that can be presented to a multi-plant planning system depends upon what planning method is used. So let's spend a little time on the planning method to be employed.

The major ways that production planning is performed in advanced planning systems is with either heuristics or optimization. While it is conceivable heuristics could be used somehow, heuristics alone would be a poor method for making choices in a multi-plant environment. This is because heuristics alone tend to work better when there are fewer alternatives to evaluate.[14] Optimization, on the other hand, does have the capability of making these complex comparisons. But that is not the end of the discussion, because the next topic would be which kind of optimizer is to be used. This is because several different types of optimizers are used for production planning.

The oldest and still the most popular is cost optimization. If cost optimization were employed in the case of multi-plant planning, then costs could be applied to different routings, causing the lower cost routing/paths to be selected over the higher cost routings/paths—as long as there was sufficient capacity to meet demand. For instance, PP/DS contains a cost optimizer (PP/DS can use either a wide variety of heuristics—which are described in the article below—a cost optimizer, or in very rare cases, Capable to Match [CTM]—an allocation/prioritization method)

http://www.scmfocus.com/sapplanning/2008/09/21/ppds-and-snp-heuristics/

However, remember that PP/DS does not have multi-plant capabilities. PT's Galaxy APS is not generally deployed with a cost optimizer (although they offer a cost optimizer as an option), but instead tends to use a duration optimizer. Therefore the objective function of the Galaxy APS optimization is to minimize the overall time of the production plan.[15]

---

[14] I say heuristics "alone" because heuristics can be combined with an optimizer to improve the planning output. This is exactly how the Galaxy APS optimizer works in fact. Officially this is referred to as optimization combined with a heuristic-based algorithm. Most supply and production planning systems offer either optimization or heuristics, but have a heuristic-based algorithm interact with the optimizer. However, a detailed analysis of the Galaxy optimization approach is tangential to the main thrust of this book, so it makes sense to not go too far afield. This may be covered in a future book because the PlanetTogether approach to optimization is quite inventive. For those that would be interested in this material, please comment at the book's website page.

http://www.scmfocus.com/scmfocuspress/production-books/the-superplant-concept/

[15] Galaxy APS also has the ability to add optimization adjustment rules, which augment the duration optimizer so it considers or weighs other factors as part of the optimization. However, to begin this analysis without it becoming complex to explain, let us assume that the straight duration optimizer in PT is used without modification.

If there were a differential between the routings in terms of their time, then Galaxy APS would choose the alternate routing/path that could be completed most quickly, as long as there was sufficient capacity on this routing/path. In many cases, moving components between factories takes longer (when there is an actual duplicate resource in the same factory) than simply passing the component to another resource in the same factory. This can be reflected in the routing duration time. To show this, let's take a look at an example.

Multi-plant Planning Example
An example or scenario can help explain how a duration optimizer within an application with multi-plant planning capability would make its decisions.

Let's say that the internal factory durations for resource B and resource D, which are in two different factories, are both 10 hours for 100 units, but the total transit time between the factories is 15 hours. The durations in the routings could be set to reflect this. So let's view the following scenario and then see how the routings durations would be set.

1. It takes 20 hours to process 100 units at resource A.

2. It takes 10 hours to process 100 units at resource B.

3. It takes 10 hours to process 100 units at resource D.

4. It takes 2 hours to move components from resource A to resource B (which are in the same factory).

5. It takes 15 hours to transfer from resource A to resource D (which are in different factories).

6. It takes 12 hours to transfer from resource B to resource E (which are in different factories).

7. It takes 10 hours to transfer from resource D to resource E (which are in different factories).

8. It takes 25 hours to process 100 units at resource E.

Therefore, the following time would be assigned to each routing.

Production and Transit Times

Resource	End Resource (if applicable)	Same Factory or Different Factory	For 100 Units	
			Processing Time in Hours	Transit Time in Hours
A			20	
B			10	
D			10	
A	B	Same		2
A	D	Different		15
B	E	Different		12
D	E	Different		10
E			25	

Sum of Hours for Routing 1	69	Hours
Sum of Hours for Routing 2	80	Hours

Resources A and E are used in both routings, so these boxes is uncolored.

If we stick to straight duration optimization, then Routing 1 would be selected as long as there were sufficient capacity at all the resources.

However, imagine that there is a delay of 18 hours on Resource B. This would not mean that the master data would have to be updated because Galaxy APS would see that a production order is already consuming resources when the order for 100 units would be scheduled to be assigned to Resource B. (There is the possibility of changing this production order manually, or having the system change the production order automatically if it were not firmed, but let us assume that the production order that is consuming Resource B when our order for 100 units would like to be scheduled on Resource B is not movable. If too many assumptions change, then we cannot see how any one factor behaves.)

Production and Transit Times

Resource	End Resource (if applicable)	Same Factory or Different Factory	For Production of 100 Units		Delay
			Processing Time in Hours	Transit Time in Hours	
A			20		
B			10		
B					18
D			10		
A	B	Same		2	
A	D	Different		15	
B	E	Different		12	
D	E	Different		10	
E			25		

Sum of Hours for Routing 1	87	Hours
Sum of Hours for Routing 2	80	Hours

*Resources A and E are used in both routings, so these boxes is uncolored.

Now we have added a delay of 18 hours to resource B. This can be seen with the new column added to the far right. This reflects the reality that resource B is consumed until 18 hours later than the time when our order for 100 units needs to be processed. Even though the software incurs more transit time, Routing 2 becomes the preferred alternative, allowing the production order to be completed 7 hours earlier.

This example shows the benefit of having multi-plant functionality. However, production and supply planning is situational or based upon a particular circumstance. What may be the best decision on some occasions is not the best decision at different times when conditions change. The less that the planning system can "flex" to different situations, the less powerful and usable it is. And this point is critical—this assessment of power that I just made is ***independent of the method***

deployed. And this gets to a very important point: planning systems tend to be evaluated based upon the sophistication of the methods that they offer—the methods being the decision logic used by the application. However, this is only one measurement of a system's power and ability to meet company requirements. It's an important criterion, but no more important than many other factors such as ease of master data update, modeling power, simulation capability, etc.[16]

To demonstrate this point I will use an example outside of supply planning and production planning, because demand planning software is such an excellent example of this exact issue.

Demand planning software is frequently graded on the sophistication of its forecasting methods (its seasonal models, its exponential smoothing model, etc.). However, the modeling capabilities, ease of adjustment, and ease of use factors are frequently ignored. I am in no way exaggerating when I say that here at SCM Focus we can do a number of things with regards to forecasting using an inexpensive demand planning application that the largest multinationals cannot do, chiefly because we test a demand planning system that grades well in multiple aspects, not simply the sophistication of its forecasting methods. Method sophistication is only one way to measure a demand planning application, or a supply chain planning application for that matter. One could have a very powerful method, but all the other areas

---

[16] The existence of a sophisticated method within a particular application does not mean that the method is in fact well-designed. It is in fact quite simple and common to take a complex method and design software in a way that makes it difficult to use the method in a beneficial manner. These types of distinctions in the design of software cannot be made by viewing a demo or speaking with software salespeople—but requires having first-hand exposure to the application. This determination can only come from testing the application and reviewing the configuration of the system. This methodology of increasing the exposure to the application is covered in detail in the SCM Focus Press book *Enterprise Software Selection: How to Pinpoint the Perfect Software Solution using Multiple Information Sources*. SCM Focus covers several sophisticated optimizers that are frequently sold as leading-edge but in fact are not and which have problems in implementation that are not simply due to lack of skill in implementing the software, or to the users not understanding the solution.

On the other hand, another optimizer of note runs into consistent problems not because the optimizer is poorly designed or because of users, but because the VPs and directors of supply chain within the implementing company frequently misunderstand that the optimizer calculates lower inventory levels because it is more intelligent than previous methods employed. This causes these companies to override the system output, negating the value of the application. Hopefully this makes clear that each situation must be reviewed individually to get to the root problem.

of the application could be weak, making it difficult to gain much value from it. Alternatively, an application could have a very good technical optimizer, but the optimizer is so difficult to configure, and the output so difficult to troubleshoot and tune, that again, it is difficult to gain much value from the application.[17] It is quite common for companies to buy sophisticated applications which they are never able to properly implement, when they would have been just as well off (and spent considerably less money) implementing far less sophisticated methods.

Multi-plant Modeling in Galaxy APS

Now that we have spent time going over the multi-plant planning environment graphically in order to grasp the this topic conceptually, let's dig into Galaxy APS to see how it is configured.

Galaxy APS has the ability to assign resources flexibly to what are referred to by PT as Capabilities, or Capabilities to Resources. A Capability is the ability to engage in an operation—to perform work. Capabilities in Galaxy APS are powerful because of both how they can be assigned to resources, how easily they can be copied to create new Capabilities (with slight adjustments) along with the fact that *"they can be created for special product attributes, such as item classes, material types of other constraints that define whether a Resource can perform a particular type of task."*[18] [19] The Capabilities Table is shown in the following screen shot:

---

[17] Interestingly, one particular product has one of the most difficult to configure optimizers as well as one of the most dated optimizer designs. Usually the product is not implemented with its optimizer because the results tend to not make any sense, it takes the most expensive hardware resources to run, and it is extremely difficult to troubleshoot and to tune. Users universally hate it—and yet it is one of the most popular production planning applications in the market! It is amazing how an application can receive the lowest grade on every single one of the criteria used by SCM Focus to evaluate software, and regardless, the software can continue to sell well in the market.

[18] PlanetTogether APS Product Training.

[19] What this means is that the Capabilities can very much depend upon a variety of factors, making them extremely flexible. *"For example, a packing Resource may have the Capabilities: pack, 8 oz, 16 oz, and 32 oz."*

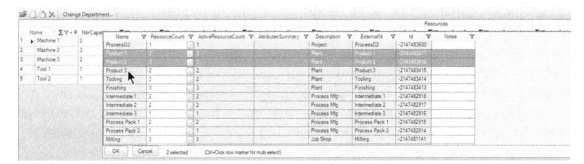

In this example, Product 1 and Product 2 Capabilities are already assigned to Machine 1 Resource.

In Galaxy APS the assignment of Capabilities and Resources is bi-directional, meaning that a Resource can also have a Capability assigned to it by opening up the Capabilities table and assigning resources in the same way. This is show in the following screen shot.[20]

---

[20] I am trying to keep the focus on superplant functionality. However, this ability to manipulate master data so easily in Galaxy APS is of great value not only during the implementation, but also for maintenance after the system is live. This means that users—those that actually know the master data the best—can take a bigger role in controlling the assignments between Capabilities and Resources without the need to rely upon technical resources. In addition to assignment, the creation of new Capacities and Resources is similarly fast and straightforward. These are important factors related to both implementability and maintainability.

Resource Configurator

Resources | Capabilities | Optimize Rules | Cells | Capacity Intervals | Recurring Capacity Intervals

#	Name		ResourceCount	ActiveResourceCount	AttributesSummary	Description	ExternalId	Id	Notes
1	Assembly	1		1		Job Shop	Assembly	-2147481132	
2	Blue	2							
3	Brown	1							
4	Coat	1							
5	Drum	1							
6	Finishing	3							
7	First Step	1							
8	Green	2							
9	Grinding	2							
10	Heat Treat	1							
11	Intermediate 1	2							
12	Intermediate 2	2							
13	Intermediate 3	1							
14	Machine Operator	2							
15	Magenta	1							
16	▸ Milling	3							
17	MRP Line 1 or Lin	2							
18	MRP Line1	1							
19	MRP Line2	1							
20	Paint	1							
21	Process Pack 1	2							
22	Process Pack 2	1							
23	ProcessA1	1							
24	ProcessA2	1							

Capabilities

Name	NbrCapabilities	PlantName	Active	AttributeCodeTableName	AttributesSummary	AutoJoinSpan	AutoSplitSpan	BatchType
Red2	1	Simulator	✓			00:00:00	00:00:00	None
Brown	1	Simulator	✓			00:00:00	00:00:00	None
Tank	1	Tanks	✓			00:00:00	00:00:00	None
Consume1	1	Tanks	✓			00:00:00	00:00:00	None
Consume2	1	Tanks	✓			00:00:00	00:00:00	None
Mill 1	1	Job Shop	✓			00:00:00	00:00:00	None
Mill 2	1	Job Shop	✓			00:00:00	00:00:00	None
CNC 1	2	Job Shop	✓			00:00:00	00:00:00	None
CNC 2	1	Job Shop	✓			00:00:00	00:00:00	None
Operator 1	1	Job Shop	✓			00:00:00	00:00:00	None
Operator 2	1	Job Shop	✓			00:00:00	00:00:00	None
Tool 1	1	Job Shop	✓			00:00:00	00:00:00	None
Tool 2	1	Job Shop	✓			00:00:00	00:00:00	None
Paint Booth	1	Job Shop	✓			00:00:00	00:00:00	None
Heat Treat	1	Job Shop	✓			00:00:00	00:00:00	None
Coat (Sub)	1	Job Shop	✓			00:00:00	00:00:00	None
Frank	2	Job Shop	✓			00:00:00	00:00:00	None
Mary	1	Job Shop	✓			00:00:00	00:00:00	None

By opening the Capabilities tab in the Resource Configurator, a list of all of the resources that have been placed into the system appears. Currently, the Milling Capability is assigned to three Resources—Mill 1, Mill 2, CNC 1. However, the Resources that are assigned to the Capability are highlighted. To assign a new Resource to the Capability, all that is necessary is to select the line item that one wishes to assign.

Up to this point we can simply assume that all capabilities and all resources that we have discussed are in a single plant. In order to use this same assignment functionality to enable multi-plant planning, all that has to change is that at least one Resource would have to be a different plant. So if in the example above, Mill 1 and Mill 2 were in two different plants, Galaxy would have what it needs to compare two alternate routings/paths.

Moving Beyond Duration Optimization

In the previous paragraphs I described the fact that Galaxy APS uses a duration optimizer. Its objective function is to minimize time. We kept to the paradigm of Galaxy APS as being solely a duration optimizer without getting into its heuristic-based algorithm because it was useful to keep the scenario simple enough in other areas. The focus was on explaining how an optimizer makes decisions with different alternative routings/paths. However, as I alluded to earlier, Galaxy APS's optimizer is not limited to duration minimization. Galaxy APS also has a heuristic-based algorithm that allows other factors to count towards the optimization results. PT calls this an ***Optimization Rule, although it could just as easily have been called an optimization adjustment***. Galaxy APS has a number of available KPIs. These KPIs are controlled with sliders that force the optimizer to weigh a broad range of factors in addition to its objective function. This approach is far more powerful than a simple cost optimizer, such as those within many production planning applications and within PP/DS. This is because PP/DS only allows adjustments to be made along a single dimension: the dimension of costs.[21] After having worked on many optimization projects, the result is clear: using only one dimension for controlling the optimizer greatly limits the control over the planning output. The following article contains a further explanation.[22]

http://www.scmfocus.com/supplyplanning/2011/07/09/what-is-your-supply-planning-optimizer-optimizing/

---

[21] PP/DS "officially" allows its optimizer to be adjusted by noncost factors, but they are minor, and my testing of both the SAP SNP and PP/DS optimizers has demonstrated that any noncost factors that are offered as options do not work reliably enough to be included in any SNP or PP/DS implementation. I have an explanation of drivers that are not related to costs at this article link:

http://www.scmfocus.com/sapplanning/2008/10/11/snp-deployment-and-fair-share/

However, the problem is that many clients thinks that these capabilities work because the SAP marketing documentation and the release notes say they do. This is also a distinction, often lost on implementing companies, that just because functionality can be made to work, does not mean that it is maintainable. It is much easier to simply add functionality so it can be crowed about in the sales process, than actually making the functionality usable over the long term. Implementing companies are not lab environments with large budgets for maintaining marginal functionality. Even the largest companies tend to allocate few financial resources to planning. A quite common feature is to over-invest in the software and implementation stage—and under-invest in the planners and in system maintenance. Unfortunately too many executive decision makers have not spent the time analyzing the history of software implementation, so without this information it is difficult to know which areas to allocate financial resources. Consulting companies will tend to coax their clients into over-investing in the implementation stage because this is where their income is primarily derived. This is a misallocation of IT resources. Furthermore, functionality that is high in maintenance will tend to fall out of use after the implementation. In fact, functionality that was at one time used to sell the software, can fall out of usage for a variety of reasons—but many of them have a common underpinning in the company's under-funding of maintenance, continuous improvement of the solution, and inadequate solution socialization. (For more on solution socialization see the following article http://www.scmfocus.com/inventoryoptimizationmultiechelon/2011/05/socializing-supply-chain-optimization/.)

[22] Inventory optimization may be one of the few exceptions to this rule, because it so effectively matches the requirements of supply planning organizations (minimizing service level for a certain level of inventory, or vice versa).

On the other hand, Galaxy APS's heuristic-based algorithm can work in conjunction with Galaxy APS's duration optimizer to adjust optimization across a wide number of dimensions. How to use an optimization rule in PT is shown in the screen shot below.

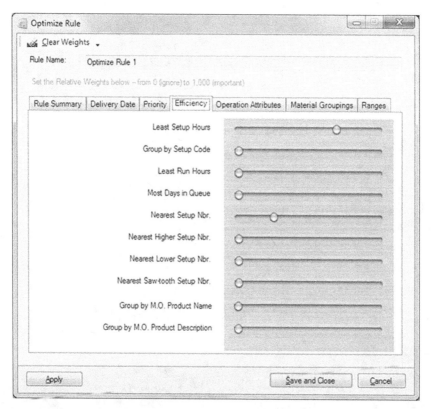

Optimization Rules—or augmentations to the optimization objective function of minimizing duration—can be added. An Optimization Rule is created by moving sliders in any of the Optimization Rule categories (Delivery Date, Priority, Efficiency, Operation Attributes, Material Groupings, and Ranges). A combination of all of the sliders makes up an Optimization Rule, which is then saved. Multiple Optimization Rules can be created and then assigned to the optimization planning run.

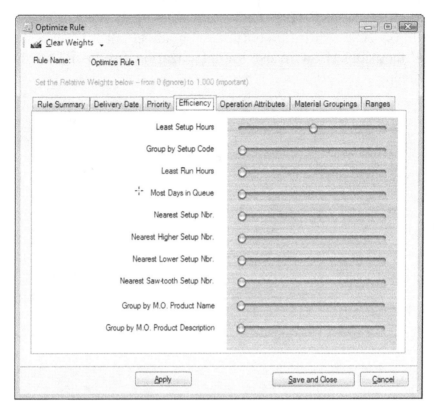

Here is a second Optimization Rule, which has sliders set on a different tab.

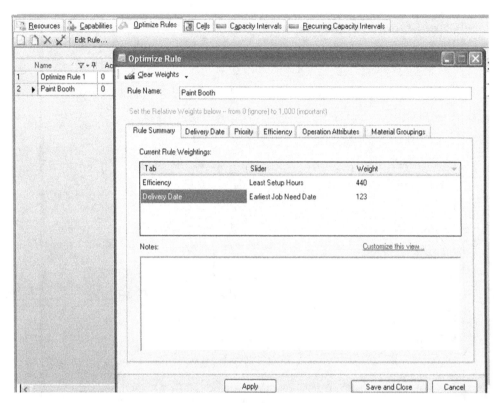

Here you can see the weights that have been applied to each Optimization Rule. You can add any number of Optimization Rules. Beyond that, not only can Optimization Rules be assigned to the overall model, but also to individual Resources as the screen shot on the next page shows.

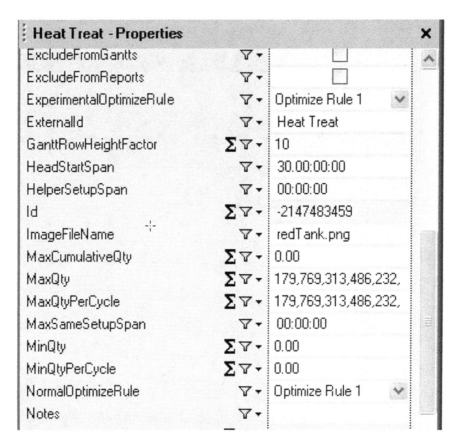

Heat Treat - Properties			×
ExcludeFromGantts	▽▾	☐	
ExcludeFromReports	▽▾	☐	
ExperimentalOptimizeRule	▽▾	Optimize Rule 1 ▾	
ExternalId	▽▾	Heat Treat	
GanttRowHeightFactor	Σ▽▾	10	
HeadStartSpan	▽▾	30.00:00:00	
HelperSetupSpan	▽▾	00:00:00	
Id	Σ▽▾	-2147483459	
ImageFileName	▽▾	redTank.png	
MaxCumulativeQty	Σ▽▾	0.00	
MaxQty	Σ▽▾	179,769,313,486,232,	
MaxQtyPerCycle	Σ▽▾	179,769,313,486,232,	
MaxSameSetupSpan	▽▾	00:00:00	
MinQty	Σ▽▾	0.00	
MinQtyPerCycle	Σ▽▾	0.00	
NormalOptimizeRule	▽▾	Optimize Rule 1 ▾	
Notes	▽▾		

For this Resource the Optimization Rule 1 is applied. Also notice that there is an Experimental Rule. Experimental Rules tend to be used during simulation.

PT's user manual has the following to say on Experimental Rules:

> *"The system has been preconfigured with a default rule that applies to all Resources. During implementation, additional rules can be set up and applied to specific Resources. Each Resource can have a normal Resource rule and an Experimental Rule. During optimization you can specify which rule you want to apply. There are several ways you can experiment with optimization rules."*

Unless a special planning run is completed that breaks out the multi-plant products and locations from the non-multi-plant products and locations (and I am by no means recommending this), the same Optimization Rule that applies to non-multi-plant products and locations applies to multi-plant products and locations, as they are all part of the same optimization planning run.

Multi-plant Planning in PlanetTogether

Now that we have reviewed how multi-plant planning works conceptually, we can dig into how to configure Galaxy APS in order to enable multi-plant functionality. Several quotations from the Galaxy APS training manual will start us off:

> *"If plants are autonomous, meaning there is no interdependency among the plants, they can be optimized separately. There is no fixed relationship between users and plants so planners can schedule plants together or individually as necessary."*

This would be the starting point in terms of configuration. Configuration typically begins with setting up the model for simpler scenarios and then moving to the more complex scenarios. In fact, Galaxy APS can be configured with any number of alternate routings/paths. But in order for the routings to be actually used, a Manufacturing Order must be available to more than one factory.

In the quotation below, PT explains how much flexibility can be provided to manufacturing orders.

> *"Each Manufacturing Order can be locked to a specific Plant by having its Locked Plant field set. If this is not set then the Optimizer will schedule operations based on Capabilities alone (as if the Resources were in the same Plant). There are also 'Can Span Plants' settings in the Manufacturing Order and in the Job to control whether the M.O. or Job can have operations scheduled in multiple Plants."*

Therefore, multi-plant planning is essentially the default setting for Galaxy APS as long as the ***alternate routings exist***. If alternate routings exist—and let's say that they span factories—then the optimizer will attempt to use the alternate routing/path if the circumstances are right for it. However, even if the alternate routing/path would be feasible, Galaxy APS will not use it for planning purposes if Manufacturing Orders *are locked to a particular factory.*

One might ask, *"Why would one want to lock a Galaxy APS Manufacturing Order to a particular factory?"* There can be several reasons for doing so. For instance, while a product could be made in multiple factories, based upon the final destination to a customer, it may make sense to have the order produced in a particular factory that is in close proximity to this customer, thus reducing lead time and shipping costs. Of course there can be other reasons, which will be explained further on in the book. However, the main point is that locking can be used selectively throughout the model and locked and nonlocked Galaxy APS Manufacturing Orders can be created in the same manufacturing run.

Multi-plant Optimization

The standard design of SAP APO and of most other supply planning and production planning applications is the following:

1. The supply planning application creates the initial supply plan and the initial production plan. If the supply plan has production resources, then either a constraint-based planning method is used or capacity-leveling. The supply plan applies for the supply planning horizon, which is frequently a year or more. The objective of incorporating constraints into the supply planning application is to pass a production plan to production planning that is feasible within each week.[23]

2. The supply plan is then passed to the production planning system, which makes adjustments to the initial production plan based upon much more detailed master data. Also, at this point the planned production jobs are moved around on some type of Gantt chart within the week and sometimes between weeks. This is all for the production planning horizon. An example of how the supply and production horizon interacts for non-multi-plant systems in SAP APO is shown in the screen shot on the next page:

---

[23] I do not spend time describing the distinctions between capacity-leveling and constraint-based planning in this book. The SCM Focus Press book *Constrained Supply and Production Planning in SAP APO* does cover this topic.

The main focus of most supply chain planning vendors that develop software in this area has been not to integrate the supply planning application and production planning application. Most software vendors simply assume them to be two different things. The weakness of this design is the natural inconsistency between the supply planning application and the production planning application. For example, SNP and PP/DS—and most other supply and production planning applications—work off of a different set of assumptions. In environments where there are dependencies between production, such as when a finished good in one factory is fed by semi-finished goods or components (or the components are in a third factory feeding the semi-finished goods plant—which I have seen at several companies), then the production planning and scheduling across the various plants ends up missing out on a number of planning opportunities that a multi-plant planning system could leverage.

Understanding the Outcomes of Supply and Production Strategies

Executives at most companies are trying to do too many things at once—things that essentially have contradictory outcomes. For example, executives want the reduced manufacturing costs of outsourced manufacturing, but also want to reduce their inventory. In general, it is difficult to meet multiple contradictory objectives but it is even more difficult when the applications selected are not appropriate for the requirement.

The Increased Complexity of the Supply and Production Business Environment

The complexity of the supply and production planning environment has increased significantly in the past three decades. Several reasons for this include the following:

1. *Outsourced Manufacturing*: The outsourcing of manufacturing to low-cost facilities in countries with long lead times from consumer markets.

2. *More Hand-offs in the Supply Network*: Increasingly complex relationships between companies and their suppliers, such as subcontracting, contract manufacturing, VMI, consignment (the transfer of inventory without the transfer of ownership), and intercompany transfer (supply planning must conform to the ownership transfer which is optimized for tax and other objectives unrelated to supply planning objectives).

3. *Growth by Acquisition*: This leads to companies (sometimes) having to combine supply chains that were previously separate.

4. *SKU Proliferation*: Marketing is very much in control of most US companies, and although most new products fail and are most often bad investments for the company, they continue to be introduced at a high rate. According to AcuPoll and Harvard professor Clayton Christensen, roughly ninety-five percent of new products fail.[24] Marketing only includes the costs of these introductions on marketing's direct costs of performing the introduction, never on the cost to supply chain operations and planning to manage all these new products.[25] New products include adjustments to products that make them better, or slight variations on an old product, but also just as importantly changes that cut out costs. Something supply chain planning has had to get used to is that no matter what the cost—changes must be a constant. Many "new" products are not recognizably new to consumers. For instance, in the spirits industry, a small change to the glass in a bottle will mean the creation of a new product number and the need to port the old demand history to the new product number. Product-location databases are in constant flux. What this means for planning is that most of the product database is "dead"—that is, planning systems are filled with product numbers that have no activity. As computerization has improved the ability to apply mathematical methods to planning, marketing has been on the other end, making the job of planning more difficult with not only continual new products, but promotions that erode the quality of demand history. Few companies are intelligent enough to buck this trend.

---

[24] Amusingly this new product failure rate has been consistent in the modern era—continuing right into the period of "Big Data." It is quite interesting that with all of these improved analytics, the percentage of new products which are successful cannot be improved even one percent by all of these amazing analytic products.

[25] Several examples of this inefficiency include reducing the law of large numbers, a major benefit to high sales items in the supply chain. The law of large numbers means reduced variability on the demand of products—leading to lower forecast error and lower safety stock. Another example is the necessity for shorter production runs, increasing the costs of the production of each produced product, as well as increasing the lead times and the lead time variability (the other factor in safety stock). Shorter production runs scheduled for more products means that each individual product must wait longer to be produced. The common trend has been to grow revenues in small increments, but to grow the product database enormously, so that the company has less average revenue per product. All of this is going on in the background, and companies are still often confused as to why they can't seem to increase their forecast accuracy.

http://www.scmfocus.com/demandplanning/2012/03/how-trader-joes-reduces-lumpy-demand/

As the article above describes, in most cases (the unusual grocery chain Trader Joes being the exception), it is exceedingly rare that supply chain planning departments have any say in SKU proliferation or in the lengthening lead times that they deal with due to the company using highly-integrated outsourced factories. Supply chain departments are expected to respond to the departments that drive strategy in the company, which in the US tends to be marketing, sales and finance.

Multi-plant Planning Versus the Common Manufacturing Trend

While multi-plant planning is required at some companies, other companies are moving in the opposite direction, away from integrated factories with increased outsourcing of subcomponents where the OEMs essentially take on the role of the general contractor. This is of course the contract manufacturing model that was covered earlier. However as both multi-plant planning and subcontracting/contract manufacturer planning are superplant functionalties, the requirement for superplant-capable software continues to rise.

Apple is an example of a company that does no production itself, and instead relies on overseas manufacturers to do all their manufacturing for them. In this model, the OEM makes almost all the profit and contract manufacturers make a margin of just a few percentage points on cost structures, which are based in the lowest-cost countries. It remains to be seen how long these companies can operate this model without becoming displaced by retailers or contract manufacturers that create their own brand, as the electronics company ASUS has done. ASUS began as a contract manufacturer, but they are now a major consumer brand as well as a contract manufacturer for other well-known brands.[26] Other familiar companies are similar. Samsung manufactures an enormous number of LCD

---

[26] Something that should be pointed out is that through contract manufacturing, an enormous amount of intellectual property (as well as process knowledge) is released outside of the companies that originally controlled this information when they performed their own production. Many companies that have engaged in substantial manufacturing outsourcing, particularly if it was to countries without intellectual property protection, can expect to see low-priced competitors that can do exactly what they do—because these contract manufacturers are currently doing it. These are the types of broader decisions that cannot be made on a simple manufacturing cost comparison basis.

screens for Apple, while competing directly against Apple's iPhone with the far less expensive Galaxy line.

However, for companies that are still vertically integrated—where the plants that provide the components are subcomponents for a finished good—the superplant concept is quite useful for creating a mental model of how the production and supply planning process needs to be designed in the associated planning systems. There are several important differences between a standard production design and the superplant design. I have discussed them throughout this chapter but wanted to list all of them in one location, which I have done below:

1. In a superplant, the BOM is distributed to multiple locations. However, the component and subcomponent BOMs are part of the overall finished goods BOM (either at the company itself, or at a company to which the planning company sells these components and subcomponents). Each BOM for the component and subcomponent are referred to as "modular BOMs." This distributed BOM applies to the multi-plant planning functionality and also to subcontracting superplant functionalities.[27]

2. In a superplant, production takes place in multiple locations, resulting in multiple production orders. The component and subcomponent production orders are triggered by distribution demand (through the supply network), rather than by dependent demand at a single location as part of the finished goods BOM.[28]

3. Synchronizing the continuous material flow between factories is critical to maintaining production efficiencies. However, the triggering of stock transfers may change depending upon plant proximity. In plants that are in close proximity to each other, the supply planning system may not need to be involved, as is described in detail in the

---

[27] Modular BOMs can be very easily created from master BOMs and version managed—if a company has the right BOM management software. However, most companies do not have this—so modularizing a BOM is much more difficult that it needs to be. The topic of BOM management systems is covered in the SCM Focus Press book *The Bill of Materials in Excel, Planning, ERP and PLM/BMMS Software.*

[28] Distribution demand is the demand that is sent between locations in a supply network. This is explained in detail in the following article: http://www.scmfocus.com/sapplanning/2012/10/15/understanding-the-flow-of-strs-and-prs-through-apo-with-a-custom-deployment-solution/

following article. http://www.scmfocus.com/sapplanning/2012/07/24/
synchronizing-integrated-factories-with-stock-transfers/

4. In a larger sense, each factory can be thought of as a work center, and
the flow through the supply network—for internal locations at least—can
be considered a routing. Of course, this mental model, which is really the
mental model of a superplant, only works for applications capable of multi-
plant planning.

5. Because production is distributed across factories, when non-multi-plant
planning software is used to plan a multi-plant planning problem, *the overall manufacturing process cannot be capacity-constrained using a
single bottleneck resource on a "single global production line."* In
effect, each production line in each factory must be separately constrained,
and *not as a single production process—which it in fact is*. With
most production planning software, when all work centers/resources are
in a single location, a single bottleneck resource can effectively constrain
the entire production process. Generally speaking, this is one of the central concepts to production planning and the reason why there is a strong
benefit to managing an entire sequence of work centers/resources *with a
single resource*.[29] This also explains why production resources are by far
the most common resource types to be incorporated into supply planning
systems, a topic which is covered in depth in the following article: http://
www.scmfocus.com/supplyplanning/2011/10/02/commonly-used-and-unused-
constraints-for-supply-planning/. However, Galaxy APS can constrain a
bottleneck/drum resource along a routing regardless of how many plants it
passes through. As would ordinarily be necessary, the lead times between
the plants would have to be added into the system—but this is a simple

---

[29] This is in fact is a problem when a direct sequence of work centers/resources does not apply. A good
example of this is with with process manufacturing. Most production planning and scheduling software
which has been developed is designed for discrete manufacturing environments—it's the largest production planning software market and the easiest problem to solve—so naturally vendors have tended to
target this market first. But, they will then often sell those same applications into other manufacturing
environments—not by adjusting the application, but by adjusting their marketing literature. Companies
choose inappropriate applications all the time, and then end up with major issues that negatively impact
their planning for years—and this is just one example of why. This topic of matching production planning and scheduling functionality to process industry requirements is covered in the SCM Focus Press
book *Process Industry Planning Software: Manufacturing Processes and Software*.

matter. Rather than having a transition of several minutes or hours between the feeding resource and the receiving resource, as would be the case if the resources were in the same plant, the lead time might be multiple hours or days when the resources are in different plants.

PlanetTogether and Planning the Superplant

Here are some important points related to Galaxy APS's multi-plant planning functionality:

1. In the Galaxy APS application, there is functionality that allows jobs to be assigned to either one plant or multiple plants. If the job is set to "Can Span Plants," each of its Manufacturing Orders is eligible to be scheduled to different plants. *"Material can flow between plants without having to issue transfer orders. This assumes that both plants are supplying the same warehouse."*[30] This means Galaxy APS can search through the network of factories and "choose" where to create the Manufacturing Order. Because of this fact, Galaxy APS can treat all of the factory locations as if they are one virtual location. Galaxy APS calculates not only the production lead times, but also the supply planning lead times between the locations when it sets up a schedule that will meet demand.

2. In the example of building chairs, Galaxy APS's Manufacturing Order—which can span plants—may start a cutting operation in one plant while the next operation, a sanding operation, might occur in another facility. The option "Can Span Plants" must be checked as "true" at the Job level in order for the Manufacturing Order setting to take effect. Galaxy APS is designed to mix and match Resources across plants in order to string

---

[30] What this means is that in order to actually move the material between locations to support the Manufacturing Orders, the stock transfers would have to be created in either APO or the SAP ERP system. If, for instance, APO were used for supply planning, the STOs could be created in APO with an interface, the load built with SNP's Transportation Load Builder (TLB), and then sent over to SAP ERP. If APO or another supply planning system is not in the mix, then the stock transfers could be generated directly in SAP ERP in order to support the combined supply and production plan generated by Galaxy APS.

together the necessary production activities in the correct sequence—regardless of where in the supply network the Resources are located. These are alternate Routings that give Galaxy APS the maximum flexibility to choose the best possible solution—given all of the constraints.

3. Similar to locking an Activity to a Resource, a Manufacturing Order can be locked to a Plant to ensure—for any reason—that the Manufacturing Order is started and finished within that plant. So while Galaxy APS can be given maximal flexibility as described in the previous point, it can also plan like an ordinary production planning system.

4. A common reason for locking a Manufacturing Order to a Plant is proximity to the customer who ordered the product. If a Manufacturing Order is locked to a Plant, every Operation belonging to the Manufacturing Order will also be locked to that Plant.

Manufacturing Order locking can be seen in the screen shot below:

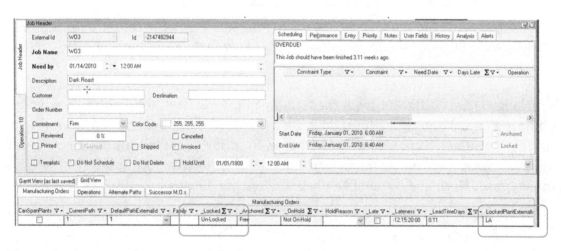

To perform cross-plant or multi-plant optimization, in addition to unlocking the Manufacturing Order you must also actively select to optimize the schedule for "All Plants" as shown in the screen shot on the next page:

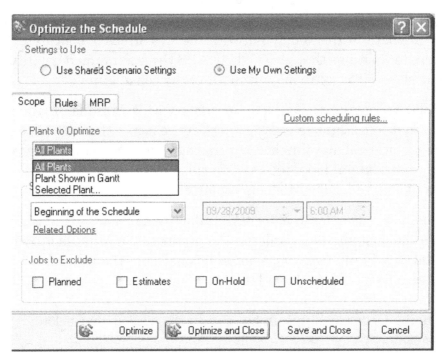

The options available in the drop-down for "Plants to Optimize" are All Plants, Plants Shown in Gantt (which means the plants that are brought up in the user interface) and Selected Plant.

This approach is different from the "supply planning first, followed by production planning" approach that I have described. When Galaxy APS is configured to do so, the Manufacturing Orders are created first, for a long planning horizon, and MRP is run off of the optimized production plan for the total planning horizon in one pass. Here are some of the benefits of this design.

1. Resources can be optimized by factory or across all factories within an Instance.

2. Each factory can be optimized according to its own optimization rules.

3. Users can use their own optimize options or share a common setup.

4. Each factory can have its own Plant Stable Span to enforce schedule stability.

5. Factories can share common resources.

6. Multiple warehouses can be defined for inventory management.

7. Inventory information can be filtered by warehouse.

8. Capable-to-promise (CTP) requests can specify which warehouses the request can draw from.

9. CTP can also be run in each factory where the item can be produced.

PlanetTogether Questions and Answers

The more one works with PT, the more unique both their technology and their implementation approach appears from other production planning and scheduling applications. In speaking with PT representatives, I found some of the following features of PT implementations to be quite interesting. I have presented these features in the form of questions and answers below:

Question #1: How long is the typical Galaxy APS production planning and detailed scheduling horizon?

Answer #1 (from PT): Most of our clients favor longer scheduling/planning horizons (between one and three months of detailed scheduling and twelve months of planning) for more visibility—especially when they have long lead-time make or buy items.

At first I found this answer surprising. I questioned why there would be such a large discrepancy between the typical production planning horizons of APO clients and PT clients. I started thinking about the fact that SAP and PT tend to sell to different types of clients. Generally speaking, SAP (and particularly SAP APO) is used at the largest companies, while PT tends to sell to smaller companies. Both vendors have customers in a wide variety of industries, and the production planning horizon and scheduling horizon should normalize around industry rather than company size. However, the answer actually relates to how Galaxy APS is used versus how PP/DS is used.

While Galaxy APS is generally marketed as a production planning and scheduling tool, it can actually perform both supply planning and production planning, depending upon the supply planning requirements.[31] So while SAP customers are implementing both SNP and PP/DS, PT customers are implementing a single product that covers both planning functions. It also means that PT customers are getting much more information about the production schedule than SAP customers because the scheduling horizon of most customers is three months. Details about how the planning horizon and related settings can be configured are covered in Appendix B: "Time Horizons in Galaxy APS."

Question #2: Do most companies that implement Galaxy APS understand that Galaxy APS's unified planning approach is a very different design than the traditional design?

Answer #2 (from PT): **I don't think they appreciate this difference.** *I think they just assume that this is how it should work—that the short-term schedule is in sync with the long-term plan. It makes sense if your mind isn't already thinking along the lines of traditional planning systems.*

PT's approach with regards to integrated supply and production planning, as well as the multi-plant functionality, is highly innovative and quite divergent from the standard software design in this area. However, one of the problems with innovative ideas is that they are more difficult to process mentally for the people and companies that could potentially benefit. In fact, when I first read Galaxy APS's documentation, I had to reread it several times, and then e-mail and discuss it with PT. This was all necessary for clarification because it was so different from the other applications I had worked with. In my mental model, the production was assigned to a single location. The only question was how much to make out of that location and when to create the planned production orders.

This led me to conclude that multi-plant functionality really needs to be made quite clear, and to be compared and contrasted with the traditional approach to production planning. Otherwise it will take quite a bit of time for companies to adopt this innovative functionality. This is of course why I decided to explain

---

[31] At least this has been the Galaxy APS marketing up to the point of the publication of this book.

Galaxy APS along with a contrasting application to show the new versus the traditional design.

Question #3: Are companies able to take advantage of Galaxy APS's ability to choose among many alternate routings and to treat all factories as if they are one large factory? Does this realization sink in later, earlier, etc.?

Answer #3 (from PT): I think most companies create their work orders to a particular factory in the ERP and don't give Galaxy APS the choice—they schedule as separate/disconnected factories. That said, we're working with a new fifty-plant client now that plans to input alternates to Galaxy APS for the planning so they can capacity-balance across factories as you're suggesting.

As I stated previously, it took me years to understand that Galaxy APS had this capability. I was aware of PT and had the Galaxy APS application on my laptop, but I simply never thought to ask about the functionality. I found Galaxy APS's capabilities because I happened to have a client with these exact requirements, and also happened to have a PT user manual and access to PT to ask questions. Companies may select PT for a variety of reasons, but they also obtain the capability to perform multi-plant planning and scheduling in addition to the other things that Galaxy APS can do. The second portion of the answer is shown below.

Answer #3 Continued (from PT): On a related note...given the high volume of data in a fifty-plant twelve-month plan, with this client we're going to take the approach of having a Galaxy APS "Instance" for planning and a separate Instance for detailed scheduling. In this case, the detailed schedules will be imported into the plan. This way the plan will take the detailed schedule into account without impacting its performance. With smaller companies this isn't necessary; they can just use one instance for both.

I found this answer interesting—primarily because I continually run into situations with SAP APO where clients try to get far too much out of a single server. This condition becomes worse on global APO implementations, which of course increases the processing load while at the same time shrinking the windows available to perform that processing. I have never been on a project where the application is split this way. But it puts Galaxy APS in interesting performance territory. It means that each server can be specialized for each task, and can be made more efficient at performing that task.

Question #4: Do companies communicate to PT that they have benefited from a more stable supply/production plan that is consistent throughout the planning horizon?

Answer #4 (from PT): I don't recall ever getting this feedback but our consultants might. I do know that in the sales process they highlight this need—for the detailed schedule to drive the early part of the long-term plan.

I am surprised by this answer. If it comes from a sequential supply planning and production planning process (driven of course by their previous application), then I would be at a loss as to how they would not observe a more stable schedule when using an integrated application like Galaxy APS.

Conclusion

As discussed, multi-plant planning, sometimes called multi-site planning, is the ability to model and make decisions to schedule production between alternate internal production locations that can produce the same product. By definition, companies that have components and subcomponents of final finished goods that are moved between factories have a multi-plant planning requirement. And this requirement applies to all manufacturing environments (discrete, repetitive, process batch, process continuous). A company that does not have multi-plant planning requirements when they start out, will have these requirements as soon as they choose to consolidate one stage of manufacturing to a single location in order to benefit from economies of scale and economies of specialization in that manufacturing process. By implementing multi-plant planning software, the company may be able to supply multiple plants. It is able to manufacture more product with fewer resources because it can receive a higher production utilization from its resources. Companies that have software that is capable of multi-plant planning are in a better position to leverage the production efficiencies of manufacturing consolidation. Therefore, software that is capable of multi-plant planning may be implemented by companies that already have multi-plant configurations, or software that can do multi-plant planning may enable a company to move to a multi-plant configuration only because they have no effective way of properly planning and controlling the factory in an alternate configuration.

Multi-plant planning is a more realistic representation of the real modeling requirements within many companies. This is because factories do not merely accept raw materials and ship finished goods. Instead, many factories receive raw materials and ship out subcomponents. Other factories receive subcomponents and ship out components or subassemblies. Many combinations of factories are possible and always have been at least to some degree. However, better communications, planning and transportation increase the opportunities to combine factories as pieces in an overall virtual production line. In environments where there are dependencies between production, such as when a finished good in one factory is fed by semi-finished goods or components (or the components are in a third factory feeding the semi-finished goods plant—which I have seen at several companies), then the production planning and scheduling across the various plants ends up missing out on a number of planning opportunities that a multi-plant planning system could leverage.

In a superplant, demand-type prioritization can sometimes be a requirement. If a company were to allow an external customer to consume capacity for a component for which there is internal demand, and for which they were constrained, then the in-process portions of the finished good would sit in inventory until the component can be produced. This would be an undesirable outcome.

Companies often require customized designs in their supply planning and production planning systems. Superplant is the distribution of what was previously one production process at one location to multiple production processes at multiple locations. Each production process at each location produces a modular BOM. The overall production process (that makes up the finished good) is then controlled in part by supply planning and in part by production planning—which are distinct and separate applications in most, but not every software vendor's suite of applications. This extra complexity brings up a host of issues ranging from what methods to use for each product location to how to control the stock movements between the related locations.

SAP development is just one example of a software vendor that chose a traditional and sequential design for their supply and production planning. This design is essentially unchanged from when advanced planning software was first implemented broadly in the mid-1990s. SAP APO, particularly the supply and production planning modules, is based upon the solution of i2 Technologies, as SAP maintained a relationship with i2 Technologies in the late 1990s. And i2 Technologies had this same sequential approach between its products Factory Planner and Supply Chain Planner. SAP's history with i2 Technologies is covered in the article below:

http://www.scmfocus.com/scmhistory/2010/07/the-history-of-apo-and-the-influence-of-i2-technologies/

This is the most common approach to production planning software design, which is currently available in the market, and is a poor fit for companies that want to move to multi-plant planning.

Galaxy APS has multi-plant planning functionality, and can search through a network of factories and "choose" where to create the planned production order. In this way, Galaxy APS can treat all of the factory locations as if they are virtually one location. Galaxy APS is designed to mix and match resources across plants in order to string together the necessary production activities in the correct sequence—regardless of where in the supply network the resources are located.

In this chapter I posed some questions to PT and presented their answers, which indicate that while PT has developed functionality, many companies may purchase Galaxy APS either without knowing the software's full capabilities, or without implementing this multi-plant functionality. This is indicative of how previous experience and familiarity drives software development and software purchases, but also of how software is implemented. There tends to be very little questioning of software design in supply chain planning. Most consultants and specialists that work in a particular application tend to simply accept the design assumptions of that application as "the natural order of things." This is not an accurate interpretation. Rather the current predominant approach to software development in any supply software category is simply a function of the decisions made by software vendors, and there are plenty of dated and inefficient software designs in all of the supply chain planning categories that SCM Focus analyzes. There are also enormous differences in how efficiently different applications can meet requirements and our research shows that this trumps all other factors in the eventual success of a selected application in a given environment.

CHAPTER 4

Single Versus Multi-pass Planning

Galaxy APS creates both the supply planning recommendations and the manufacturing recommendations in a "single pass" or a single optimized planning run. Galaxy APS can either run MRP after the optimization—which would be two-step or multi-pass planning. But it can also run MRP during the optimization. The setting for controlling this is shown in the following screen shot.

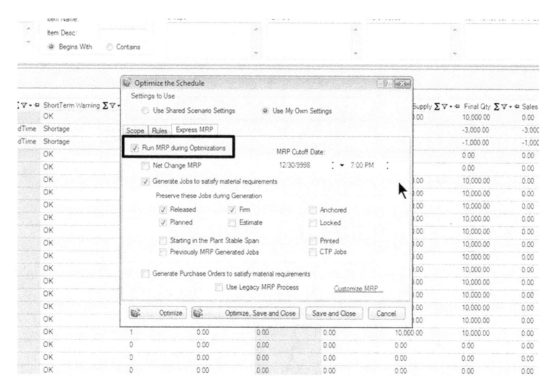

Galaxy APS uses MRP to actually create the purchase requisitions and stock transfers—but this is based upon the optimizer output and can be set as part of a single planning run.

This approach can be superior to the sequential approach (where supply planning releases its results to production planning) in several ways. First, a combined run can allow an application to do things like multi-plant planning (such as compare all the alternate paths, including those that span a plant). Secondly, the ***supply plan is accurate right off the bat***, and does not have to be passed to a production planning tool for adjustment. This eliminates disconnects between the supply planning system and the production planning system, which are described in the article below:

> http://www.scmfocus.com/sapplanning/2013/07/16/disconnection-points-between-snp-and-ppds/

Both supply and production planning can be performed in a single application and in a single pass, which means that both supply and production planning could have the same assumptions. For instance, there are no planning horizon differences between the supply planning and production planning system. Having worked a large number of combined supply and production planning implementations using two systems, I can say that coordinating just the planning horizon and planning bucket (the duration for which the planning data is stored) configuration between these two systems is a challenge. In fact, a good part of the SCM Focus Press book *Planning Horizons, Calendars and Timings in APO,* covers this topic of timing integration between a supply planning and a production planning system.

There are, however, trade-offs between two-pass and single-pass planning. Which approach is a better fit depends upon the objectives of the specific client. Most supply and production applications do not provide the option of either a single-pass or sequential processing, so in most cases the issue cannot even be brought up.

Sequential Processing Versus Single-pass

Characteristics of the Planning	Sequential Processing	Single Pass
Production and Supply Planning are Performed in a Single Planning Run	N	Y
Creates Synchronized Multi-plant Plans?	N	Y
Supply Plan and Production Plan are Passed Between Systems	Y	N
Simpler to Manage	N	Y
Effective When Set-up Times are Important in Scheduling	N	Y

This matrix encapsulates the advantages and disadvantages of single-pass versus sequential processing.

Having two different systems—one specialized supply planning and one specialized production planning, at least the way it's done in APO—is really a better fit for larger companies with more resources. Also, SNP and PP/DS work very differently and require different consultants—with different designs, user interfaces, etc.... It's quite a bit more overhead. In fact either SNP or PP/DS by themselves are more effort to maintain than Galaxy APS, so the resource commitment is in a completely different category. However if more advanced supply planning requirements exist, such as multi-source planning, then a separate supply planning system would have be connected to Galaxy APS.[32] But, for companies that do not have such requirements, performing both supply and production planning in Galaxy APS makes a lot of sense.

"Supply Planning First" Two-pass Approach

In the sequential "supply planning first design," cross-plant interactions are only observed and planned in the supply planning system—in this case SNP. The supply plan is passed to PP/DS. PP/DS then creates planned orders from the first time bucket in the planning horizon out until the end of the production planning horizon. While this production planning horizon can be set to any value, in practice, it tends to be a matter of weeks. As such, in this design the supply planning system is in control of most of the combined supply and production planning horizon, and essentially gets first crack at processing the demand. Remember that most companies will have SNP use production resources, so SNP is creating the initial production plan. However, PP/DS will adjust the initial production plan provided by SNP for what is generally two to four weeks of the PP/DS production horizon, and detailed scheduling is performed down to the level of detail of the hour.

I will cover the aspect of the planning horizon integration in more detail further on the book. While both SNP and PP/DS use the same production resources, they use them in a different manner. These differences are primarily related to the time nature of the resource. Therefore APO production resources have fields that are used by SNP and others that are used by PP/DS. Both SNP and PP/DS create planned orders along the planning horizon; however SNP creates them first, as SNP looks out much further than PP/DS. The degree to which PP/DS can adjust planned orders as they fall into the production planning horizon depends upon whether or not the planned orders have been firmed. See this article on firming and fixing in SAP APO.

http://www.scmfocus.com/sapplanning/2012/06/22/firming-in-apo/

---

[32] As will be covered in Chapter 5: "Multi-source Planning," SNP would not be sufficiently reliable or sufficiently low in maintenance for use in multi-source planning.

This is shown in the graphic below:

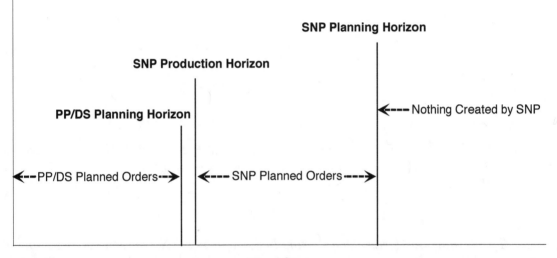

The SNP Production Horizon and
SNP and PP/DS Planning Horizons

Here we can see how the SNP Production Horizon controls whether or not SNP creates planned orders. This graphic shows no overlap between the PP/DS Planning Horizon and the SNP Production Horizon; however, this is just one of a number of different scenarios. On projects, there often is some type of overlap between SNP and PP/DS, meaning that there is a section of the planning horizon for which both SNP and PP/DS create planned orders.

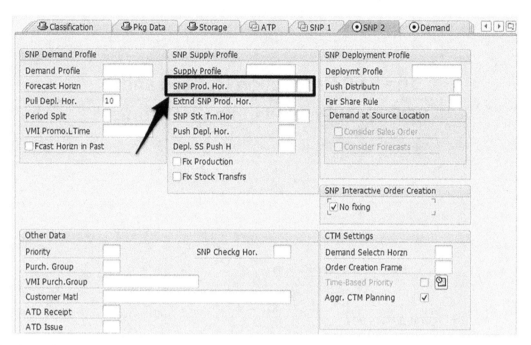

The SNP Production Horizon is shown above on the SNP 2 Tab in the Product Location Master. I have included a description of the fields in this master data area, which are related to timing.

1. SNP Planning Horizon: How far out SNP processes the supply network.

2. SNP Production Horizon: The period inside of which SNP does not create planned orders.

3. PP/DS Production Planning Horizon: How far out into the future the production plan processes for the factories. The production planning horizon is sometimes set per factory. It is set in the production planning system, but interacts with the supply planning horizon, in particular when planned production orders are created by the supply planning system versus the production planning system. In this field, you specify a location-product-specific PP/DS horizon. The production planning horizon can be inherited from the SNP Production Horizon by entering no value in the PP/DS Horizon field. If you do not enter any value for the PP/DS horizon, or if you enter the duration "0", the system automatically uses the SNP production horizon as the PP/DS horizon. Therefore, the PP/DS horizon is as long as the SNP production horizon. The graphic that appears below this field section shows

the SNP Production Horizon and the PP/DS Production Horizon. The following lists the implications of using the SNP production horizon as the PP/DS horizon:

a. The planning intervals for SNP and PP/DS are sequenced with no gaps or overlaps.

b. If the SNP production horizon also has a duration of "0", the system uses the PP/DS horizon from the planning version. You can use the PP/DS firming horizon within the PP/DS horizon to firm (short-term) planning for planning with procurement planning heuristics. SNP is permitted to plan outside of the SNP production horizon only. If the SNP production horizon is smaller than the PP/DS horizon, the planning horizons of SNP and PP/DS overlap. SNP and PP/DS can both use this overlapping period for planning.

This brief overview discussed just a few of the timings in SNP and PP/DS and was probably a bit exhausting for some readers. These timings must be synchronized between the two modules in order to make the applications work together effectively. I should also point out that one does not need to choose a single set of timings, meaning that different horizons can be used for different locations and for different products. Doing so requires evaluating the overall product and location database in order to make these timing decisions, as is covered in great detail in my book *Planning Horizons, Calendars and Timings in SAP APO*. However, Galaxy APS has no timing integration needs and no book to read on this topic because the timings are easily set in a single application.

Single-pass Planning with PlanetTogether

Galaxy APS can combine supply and production planning into a single run. The steps are described below:

1. The user clicks "Optimize."

2. The plan and schedule are created from independent demands all the way down through production scheduling and purchasing in one shot.

Galaxy APS does not currently create stock transfers automatically but will eventually. They can be created interactively at any time. As PT develops enhanced supply planning functionality, an open question is how to configure the functionality, which depends upon the specific requirement. PT is going to offer several alternatives rather than a single approach.

1. The single-pass approach will do a good job of creating schedules that deliver on time when sequence-dependent set-up times or other complex scheduling constraints are the driving decision point.

2. A multi-pass approach will do a better job at coming up with a plan that optimizes a stated objective function, but the details of the production constraints it is using will be less accurate. This could be advantageous in cases where distribution time and costs are more important than capacity details and on-time delivery.

At SCM Focus we recommend software that can be used in either a single-pass or multi-pass approach. This allows the software to be run in different ways in different planning runs.

CHAPTER 5

Multi-source Planning

One major reason companies select cost optimization for their supply planning implementations is to perform constraint-based planning. Constraint-based planning is the ability to restrict capacity, primarily in the production resources, although hypothetically companies either project or are told that they will constrain on the basis of other supply chain constraints such as transportation and warehousing.

However, another powerful motivation for selecting cost optimization is to perform multi-source planning, also known as multi-sourcing. When there is more than one source and the sourcing is determined by the supply planning application, the functionality is referred to as multi-source planning. Multi-source planning is the ability for a supply planning system to choose intelligently between alternate sources of supply. A source of supply can be a supplier, or another internal location in the supply network.

One common reason for multi-sourcing is to move to a second or even third location in order to satisfy the demand when the primary location cannot handle the capacity of the order. Another reason can be to spread, by rough percentages, the total demand among various supply sources. Although we have already reviewed different areas where multiple sources of supply can be selected by SNP, for whatever reason the term multi-sourcing refers to choosing from among multiple sources of supply for either a stock transfer or a procurement movement.

In theory, there are two methods for performing multi-sourcing in SNP. One is CTM and the other is the SNP Optimizer. How each method works with multi-sourcing is described below.

Multi-sourcing with SAP SNP Capable to Match (CTM)

In addition to being able to pick from different sources, CTM can choose from among different alternative resources. Therefore, CTM is allowed to switch to a secondary resource/PPM/PDS if the first one is consumed. However, that is a selection between resources within one location. If there are two production lines that make the same product in two different factories, CTM cannot switch to a different factory, and cannot change the source of supply based upon a resource becoming consumed in the factory with the resource set to the top priority.

Multi-sourcing with the SAP SNP Cost Optimizer

The SNP optimizer bases multi-sourcing upon the relative costs. At least it works that way in theory. I say this because I have never seen a company that turned on the multi-sourcing functionality for the SNP Optimizer continue to use it into production, but I have seen a number of companies that have tried. Interestingly, what they found was that the extra processing time for the optimizer was quite high, actually too high to meet their timing requirements for the overall solution. In the perfect scenario, one location would have a higher cost to supply than a second location.

When the primary sourcing location runs out of capacity, the optimizer, in theory, will then move to the secondary source of supply, without the planner having to do anything. The diagram below can be used to help understand this:

Multi-source Supply Planning

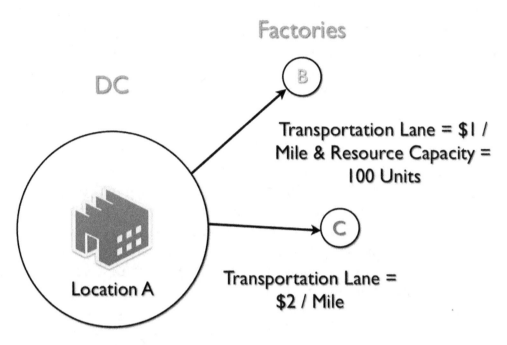

In this scenario, two producing locations have been set up as sources of supply for Location A, which is a DC. If the requirements are within Location B's capacity, Location B fulfills the requirements from Location A, because the Transportation Lane cost is only $1 per mile, versus $2 per mile with Location C. When the costs are set up in this way, nothing further needs to be done. The system will naturally source from Location B. However, if in any one time period the requirements are higher than 100 units, Location C will begin to serve as a source of supply to Location A. If the resource that produces the product for Location A goes down for maintenance, the resource has no capacity in Location B, and C becomes the sole source of supply for this material to Location A. Executive decision makers generally love that the system auto-adjusts, as shown in this hypothetical example, and foresee great cost savings from such a system. However, the reality of what occurs with multi-sourcing is quite a bit different, as explained in the following article:

http://www.scmfocus.com/sapplanning/2012/07/06/does-ctm-support-multisourcing/

In order to perform multi-sourcing with SAP SNP, a company must not only maintain the master data effectively for the multi-source option, but must also spend on the servers to make the multi-sourcing model run effectively within the available windows for processing. And the expense of testing and adjusting the configuration along with adding hardware, as well as recursively making adjustments to the multi-sourcing configuration as other adjustments are made in the system, will cost more in man hours than the actual hardware. It will also mean more difficulty in troubleshooting because multi-sourcing will be another factor which must be analyzed when an incorrect source of supply planning decision is made.

In short, multi-sourcing in APO is expensive to do. If companies are not willing to support this expensive solution, it makes little sense to head down this path. Multi-source planning in SNP is not some simple functionality that can be activated without a serious resource commitment. Right now, quite a few companies plan to turn on multi-sourcing in supply planning applications that have a strong likelihood of never working properly. None of the consulting companies document the problems with multi-sourcing because: a) They are not in the business of providing free information, and b) They do not want to adversely impact their relationship with SAP. Therefore the same mistakes will be repeated.

However, the hypothesis of how multi-sourcing works can still be used and applied to a software selection of applications that are capable of multi-source planning. Such an application should really be straightforward to configure. Each product should be able to have priorities set for it per location combination so that the optimizer can easily follow the prioritization sequence from high to low. Furthermore, the optimizer must be able to process the multi-source-enabled supply network without an undue processing time penalty. Processing time is a known issue with this functionality, and the issue cannot be proven one way or another during a simple demonstration. Therefore, some benchmarking using a significant number of products and locations is necessary in order to know if the software vendor's optimizer can really efficiently process this added alternative decision analysis along with all the other things the optimizer has to do during a planning run.

Sources of Supply in PP/DS

This book is focused on sources of supply for SNP. However, it is worth noting that PP/DS uses the same approaches as SNP to determine sources of supply.

For instance, its optimizer can also select from PPM/PDSs with different costs, can recognize quota arrangements, etc.

Conclusion

Multi-source planning allows a company to choose the best possible option, without manual intervention. This means that potentially, a better sourcing decision will go unrecognized without this functionality. Without multi-source planning functionality, multiple sources can be determined by adding quota arrangements or priorities between the locations, but these things are higher in maintenance and do not add the degree of flexibility that is desired by many companies. However, with a properly functioning multi-source planning system, the application will be able to flex to the changing circumstances, and do so as part of the normal planning run. Multi-source planning is a very close parallel to both multi-plant and subcontract/contract manufacturing planning in that all three functionalities allow the planning system to flexibly select a location that is the best decision among a series of alternatives and can adaptively switch between these alternatives as the circumstances change along the planning horizon.

CHAPTER 6

Subcontracting Planning and Execution

In Chapter 3: "Multi-plant Planning," the following graphic about how Galaxy APS treats the subcontractor location was shown. It explains how subcontracting planning is part of superplant functionality. Essentially subcontracting is the same concept except a different company owns the subcontractor (more on how it is different is coming up shortly). Galaxy APS can treat the subcontractor location in the similar way that it treats duplicate manufacturing capabilities that are in different plants. This is shown in the following screen shot.

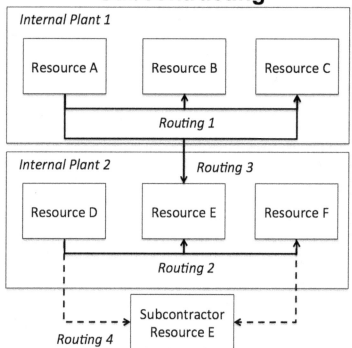

Background on Supply Chain Subcontracting

The term "supply chain subcontracting" is adopted from the general term of "subcontracting," where a subcontractor (also known as a subcontractee) provides services to a contractor or general contractor in order to fulfill a contract. Within supply chain management subcontracting has a very specific definition: Subcontracting is when a portion of the overall manufacturing process is assigned to an external company (called the subcontractor or subcontractee).

How does this differ from simple procurement? If a company (called the contractor) purchases a component and this component was manufactured by the supplier/subcontractor, didn't the purchasing company "subcontract" to the supplier? Well, subcontracting is when the purchasing company/contractor provides a component as an input raw material or semi-finished good to an external party, which then performs a manufacturing process for the contractor. This is different than contract manufacturing, where the contracting manufacturing company manufactures the entire finished good of the purchasing company.[33]

Interestingly, while the term is used widely in industry, subcontracting (for supply chain) has no official definition on the Internet, aside from the definition that I created after I found out that it was lacking.

http://www.scmfocus.com/productionplanningandscheduling/2013/07/15/subcontracting-supply-chain-definition/

In the following situations, the companies are engaging in subcontracting:

- A company sends out products to be packaged.

- An OEM provides some of the raw material to the subcontractor that manufactures the entire product.

- A company sends out a partially manufactured product, and then sends out the items to another company for a specialized manufacturing process, and then has the products returned to it in order to complete the manufacturing process.

There are many possible process flows for subcontracting. Some have the products returning to the OEM for further processing and some have the products being completed by the subcontractor and then shipped from the subcontractor. The multiple subcontracting processes place an extra layer of complexity on planning and accounting for subcontracting in a supply planning system. However, because of the significant complexity of subcontracting, and the fact that subcontracting could fill a book by itself, I will only discuss the most basic type of subcontracting in this book.

---

[33] *"In a contract manufacturing business model, the hiring firm approaches the contract manufacturer with a design or formula. The contract manufacturer will quote the parts based on processes, labor, tooling, and material costs. Typically a hiring firm will request quotes from multiple CMs. After the bidding process is complete, the hiring firm will select a source, and then, for the agreed-upon price, the CM acts as the hiring firm's factory, producing and shipping units of the design on behalf of the hiring firm."*—Wikipedia

Choosing Among Subcontractors

A company may use several alternative subcontractors for the same part, and require the planning system to make a selection among the various subcontractors. Just as an alternate path may be set up in Galaxy APS that strings together one or more subcontractors in a routing, multiple alternate paths may be set up that also allow Galaxy APS to choose among redundant subcontractors for the same part. In addition to selecting among various subcontractors, the planner can easily switch to an optional contractor option—if their internal resources are busy. Galaxy APS can also decide to use the contractor under the same conditions. Two different subcontracting alternate paths or routings are shown in the following graphic.

Subcontract Planning

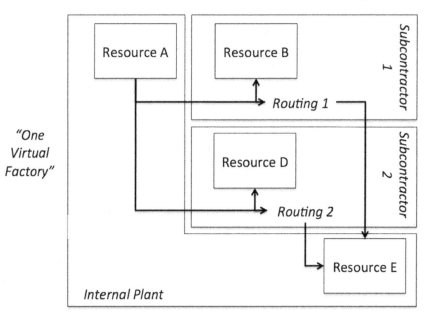

Resources can be set to finite, but more often the subcontract resources are set to infinite. This gets into another topic—getting capacity information from subcontractors, which was addressed briefly in Chapter 1: "Introduction."

Getting Parts to the Subcontractor Location

The parts that are input products to what the subcontractor manufactures may be sent from the OEM facility or from their suppliers.

- If the part is sent from the OEM location, it will, depending upon the software vendor, sometimes be modeled by the supply planning system as a stock transfer. This stock transfer is a problem, because the subcontractor location is not a true internal location.

- If the part is sent from one of the OEM's suppliers, then a purchase order is sent to the supplier from the OEM, but with a delivery location to the subcontractor address and not to the OEM location. This is relatively smooth. The real problem is the first example, when the part must come from the OEM location.

A major feature of planning systems that support a subcontract process is that they have a "switch" that allows them to change how the subcontractor location is treated depending upon the step in the process. On the other hand, this switch can also become a maintenance issue—and this complexity is one very good reason why subcontracting is rarely implemented effectively in external planning systems.

> *"Another point perhaps worth noting is that the subcontract operation(s) may have predecessor operations in the routing, thus making this a bit different from just buying components. The predecessors will have to be finished before the parts can be sent out versus a component that can be purchased at any time in advance."*
> — Jim Cerra, President and CEO of PlanetTogether

Integrating the Production Planning System with the ERP System for Subcontracting

When integrating Galaxy APS with the ERP system, the workflow to the ERP system is as follows:

Subcontracting Integration (ERP & Production Planning System)

Step	Supply or Production	Action	Process Description	For Subcontract Product	From System	To System
	P	**Setup**	Subcontractor Operation Master Data	N/A	N/A	N/A
	U	**Update**	Subcontractor Resource Master Data	N/A	N/A	N/A
1	P	Perform	Scheduling & Subcontract Operation Movement	Output	Galaxy APS	N/A
2	S	Create	Purchase Order /or Buy Direct Material	Output	Created in ERP from Galaxy APS signal.	
3	P	Sent	Feedback to Schedule from Subcontractor (PO receipt date is treated as a constraint by Galaxy APS)	Output	"Subcontract System"	Galaxy APS
4	P	Create	Completion of Purchased Order	Output	ERP	Galaxy APS

After the purchase order for the input material is created, a dependent requirement is created for the input product, which is sent from the OEM to the subcontractor. As with the purchase order, this stock transfer can also be created by either the ERP system or by Galaxy APS through running MRP, which can be part of the main Galaxy APS planning run.

In Galaxy APS, subcontractor manufacturing planning and scheduling are managed by something called an outside operation. These are the options.

1. A single resource can be created for all subcontractors, or each subcontractor can have multiple resources.

2. A subcontractor resource can be set up like any internal resource in Galaxy APS.

3. Capacity intervals can be created for subcontractor resources, just as for other resources.

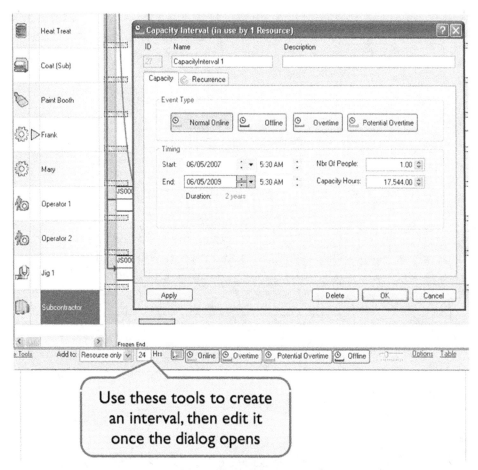

Use these tools to create
an interval, then edit it
once the dialog opens

Once the Capacity and Resources have been defined, they can be used for scheduling. Here we can see that a Subcontracting Operation is available for scheduling.

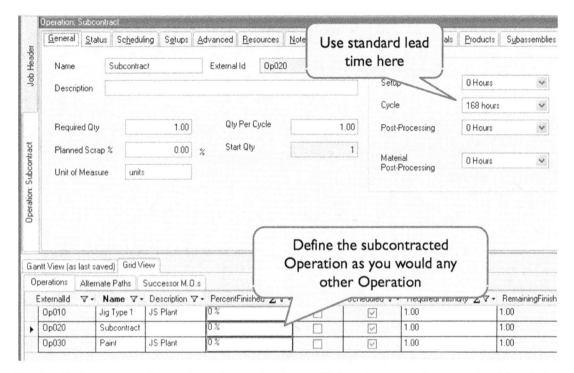

On the Subcontract Operation, the cycle time will be whatever the standard lead-time is, in this case 168 hours.

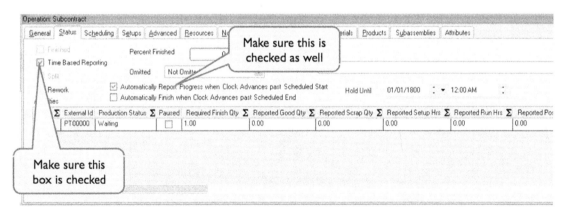

Subcontract Operations tend to be configured differently than Operations in internal plants. With an internal plant, the scheduling updates are constantly updated to the external planning application—and as time passes unless progress is reported. However, *with a subcontractor,* **this type of update usually does not occur.** *Therefore, the setting "Automatically Report Progress when Clock Advances past Scheduled Start," is activated.*

Typically one wants to also activate *time-based reporting in Galaxy APS*, which means that the system looks at run time to calculate how much time remains on an activity instead of looking at an activity.

Contract manufacturers typically provide the finished good, therefore contract manufacturing tends to be viewed as a sourcing operation. This means it is viewed as a supply planning issue rather than a production planning issue. However, this will often depend on several factors. First, a contract manufacturer can provide a component rather than providing a finished good. Secondly, the OEM may want to perform the production planning of the contract manufacturing location—and therefore the CM location may be set up as a plant in Galaxy APS and have its capacity modeled just as with a subcontractor location.

Adding an External Supply Planning System

What is described here is not, of course, the only possible combination of systems. This is just one possible solution design. For instance, the subcontractor needs a way to update their capacity information. I do not cover that topic up to this point. Secondly, I only show an ERP system and production planning system; however, an external supply planning system can also be added.

Conclusion

Subcontracting planning allows a company to use a subcontractor when their internal capacity is either insufficient to cover the demand or if other factors that are used in the optimization simply make it more effective to use subcontracting. Subcontracting planning can also allow a company to select from multiple subcontractors. Better subcontractor planning means more foresight as to when a particular subcontractor will be needed, and this improves the relationship with the subcontract manufacturer and increases the likelihood that the subcontractor will actually have capacity. Not only can these capabilities result in the obvious financial benefits such as a lower cost of acquisition for components, but can also mean less obvious financial benefits such as the ability to better meet demand.

CHAPTER 7

Combining All Three Superplant Functionalities

Up to this point we have discussed each superplant functionality independently. However, each method can be combined if software that includes all three functionalities is used. When this occurs, the company receives the full benefit of a location-adaptive supply and production planning system that can make trade-offs among various alternatives. The following graphic demonstrates how this would work.

Production Alternatives

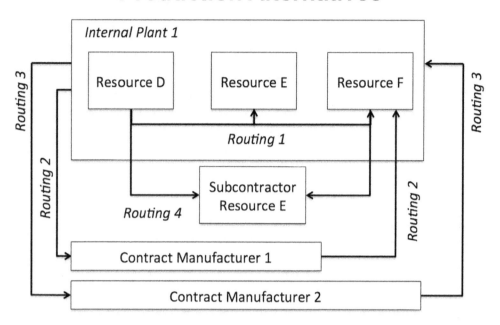

Galaxy APS can evaluate all of the alternatives that are listed above. Galaxy APS compares multiple sources of supply (not pictured), which could be brought into Plant 1, pictured above. Notice that one contract manufacturer provides an intermediate product, while the other provides a finished good. Galaxy APS would have no problem comparing each alternative to internal production.

There are too many alternatives to compare for a book graphic to illustrate. For this reason I have created the table on the next page, which shows a sample of what Galaxy APS can compare in a single planning run. The manufacturing process used in the table processes apples. It has three operations: peeling, slicing and packaging. The company buys apples from farmers and has multiple sources of supply for apples.

Compared Alternatives

Description	State of the Apples				Routing #	Product Type	Superplant Functionality
	RA	SA	SLA	PA			
Internal Plant 1				X	1	Finished Good	Multi-plant Planning
Internal Plant 2				X	5	Finished Good	Multi-plant Planning
Subcontracting for the Operations on Resource E			X		4	Semi-finished / Component	Subcontracting
Contract Manufacturer 1			X		2	Finished Good	Multi-plant Planning
Contract Manufacturer 2				X	3	Semi-finished / Component	Multi-plant Planning
Supplier (Farmer) A	X				N/A	Raw Material	Multi-sourcing
Supplier (Farmer) B	X				N/A	Raw Material	Multi-sourcing
Supplier (Farmer) C	X				N/A	Raw Material	Multi-sourcing

This matrix shows important details of the comparison that would take place each time a planning run were performed. The apples have four possible states that declare their completeness through the production process, and these are shown above with codes to save on space: RA for Raw Apples, SA for Skinned Apples, SLA for Sliced Apples, and PA for Packaged Apples. Each of the alternatives can be viewed simply as offering a different state of progress of the apples in return for a certain cost, capacity, duration, etc. With all of the alternatives entered into the system, combined with the weights for various optimization criteria, Galaxy APS can make the best choice based upon the characteristics of the alternatives versus the optimization rules.

The final three line items are the selection among suppliers. If multi-source planning functionality were enabled in a supply planning application (not in Galaxy APS as it lacks multi-source planning functionality), then all three functionalities of super plant planning would be enabled.

This is a small example of the alternatives that can be placed into Galaxy APS. On a real product, one can imagine a *lengthy spreadsheet that shows all the alternatives* that Galaxy APS computes per planning run. In fact, it makes sense to create a spreadsheet like this in order to explain how the system will make these decisions for an implementation, and as a precursor to making changes to the master data for each of the alternatives. Too often systems are implemented without any external explanation of the decisions that the system is making. Spreadsheets like this are excellent at getting all interested parties on the same page. More columns can be added to provide more detail on each of the alternatives.

It should also be noted that the optimization will not necessarily choose just one of the alternatives that move the apples into a particular state. It can select one alternative up until the capacity is exhausted, and then switch to another alternative in order to access more capacity. All of the alternatives can have minimum lot sizes, and Galaxy APS can see this information as it switches between the various alternatives.

How Contract Manufacturing Fits into Superplant

At first glance contract manufacturing can seem peculiar; contract manufacturing may look like purchasing because the OEM/contractor often receives a finished product. However, *contract manufacturing is not always simply a supply planning issue*, and this is why the contract manufacturer (CM) is managed with an alternate routing/path and does not use multi-source planning functionality, but instead may use subcontracting planning functionality.

Let's now dig into how contract manufacturing works from the business process perspective and then it can be easily understood why this arrangement is essentially a slight modification of a subcontracting relationship.

First, the CM builds to a specification. The CM may offer very similar versions of the same product to other OEMs. Let's take laptops as an example. Many laptops for different OEMs are produced off of the same production line, with only cosmetic differences between the production outputs. However, in the example of the iPhone, when Apple hires the CM Foxconn, Foxconn may not offer rebadged iPhones to other companies. Apple signs an exclusive arrangement for iPhones with Foxconn. These are examples of contract manufacturing arrangements for finished products. However, contract manufacturing also applies to components.[34]

Contract manufacturing is also not (technically, although there can be some blurring of terminology in conversation) subcontracting, because the OEM/contractor does not provide an input product to this manufacturing entity. The following transaction workflow applies.

1. The supply and production planning system models the production plant as an internal location.

2. The planning system plans the production/manufacturing order, and because it is an internal location, the transfer of the output material is sent between two "internal" locations (one the CM's location and one true internal location) as a stock transfer.

3. When the stock transfer is sent to the ERP system, the stock transfer is created, but it is a stock transfer with a billing document. This is a very common transaction type in ERP systems. This type of transaction was actually first designed for intercompany transfers and will be discussed again in Chapter 8: "The Superplant and the Integration Between ERP and the External Planning System."

A company can decide how much detail they want to model in their CM. They can set up the entire CM manufacturing process as a single operation on a single resource, or they can model actual resources within the CM. However, this will tend to be useful only if:

1. The OEM/contractor is actively scheduling the CM.

2. The CM is providing internal factory updates to the OEM/contractor.

---

[34] I combine job production and contract manufacturing into just "contract manufacturing," even though they are in fact slightly different. I do this because I believe most readers will not be interested in this level of detail. However, for those that are, I have included the following quotation on job production. *"Job production is, in essence, manufacturing on a contract basis, and thus it forms a subset of the larger field of contract manufacturing. But the latter field also includes, in addition to jobbing, a higher level of outsourcing in which a product-line-owning company entrusts its entire production to a contractor, rather than just outsourcing parts of it."*—Wikipedia

However a CM may be added with an extra routing to Galaxy APS, as if it is just another alternative. Galaxy APS can then compare all of the routings in order to select the best, depending upon the particular circumstances of the planning time interval and the criteria of the optimization. The criteria of the optimization can be finely controlled in Galaxy APS with the application of Optimization Rules. Any combination of key performance indicators can be saved as an Optimization Rule and applied to the optimizer. This adjusts the optimizer, allowing it to meet multiple objectives, but without the complexity of actually changing the objective function of the optimizer.

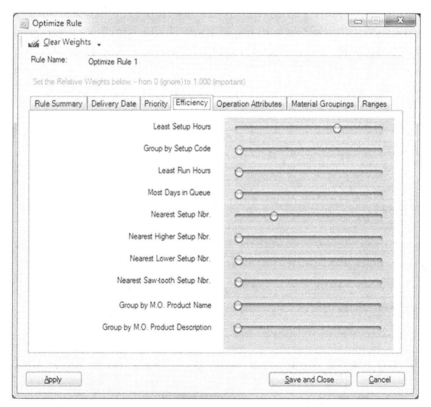

Galaxy APS uses optimization rules that adjust the duration-based optimizer for different criteria. The selection of the alternatives may change depending upon capacity issues, or it may change when the optimization rules are adjusted. For instance, if one method is less expensive but faster, the optimizer may switch to the less expensive alternative if the optimization rule is adjusted to increase the weight placed upon expense.

Understanding Optimality

Often, people will use the term "optimal" to describe a solution. However, ***what is optimal depends upon the criteria used***. Far too often optimization is thought of as a single best output that is better than any other conceivable output—that is a complete oversimplification of how optimization works in reality. Optimality quickly changes depending upon which criteria a company decides to weigh. This of course depends upon what application is being used, as well as how it is configured. With most applications there is little choice with regards to adjusting the objective function, and in most cases in supply chain planning the optimization engine can only focus on costs (with slight adjustments here and there). Galaxy APS on the other hand can adjust the optimal solution depending upon what is important to the company, and this can change from planning run to planning run. It is a simple configuration to change, but would require testing/simulation to choose the degree to which the Optimization Rules should be changed per KPI. Different runs with different Optimization Rules can be saved as simulation versions in Galaxy APS. Adjust just one Optimization Rule even slightly and a new solution is optimal. It really all depends upon the goals that are placed in the optimizer.

Superplant Planning and Processing Times

What should be clear by this point is that including superplant functionality into any application will mean that the application will have to analyze many more alternatives than an application that does not have this capability. This also means more processing time—all other things being equal. However, not all other things are equal. Some vendors have put a great deal of effort and thought into minimizing the amount of time that their optimizers take to run, while others have not. They may have created the optimizer to meet basic functioning, and then moved on to other development work.

While there is no entity that compares the performance of enterprise supply chain planning applications, a good indicator of how well the solution has been optimized for runtime performance is their hardware specification. If one application has much higher hardware requirements than another that is essentially processing the same problem, then there is a good possibility that not much effort has been spent in optimizing the performance of the procedure. I have not observed high hardware specifications coming from PT. In fact the relatively small Galaxy APS hardware sizings I have reviewed, when contrasted to the SAP applications that I am most familiar with, have been a surprise to me. This is a testament to the Galaxy APS optimizer design.

This is not the first time SCM Focus has written on substantial differences between vendors in processing time for something that seems as if it should add to processing time. For example, the demand planning and supply planning vendor Demand Works has developed an intelligent approach to managing attributes that are assigned to items to be forecasted. While some applications require significant build times to create the relationships, Demand Works does not. At first I thought that Demand Works was using a completely different technology than the competition—but they were not. However, how they had developed around this technology was simply better. As a consequence, Demand Works' software has been tested by SCM Focus and has been shown to be extremely fast and effective at creating relationships using small hardware specifications.

In software, much of how well things work comes down to design, and the design of various functions is by no means similar among various applications in the same software category.

Conclusion

The main thing to take away from this short chapter is that any of the superplant functionalities may be implemented alone, or all two or all three of the superplant functionalities may be implemented. The more of the superplant functionalities that are implemented, the more adaptive the planning output will become. However, the requirements for each superplant functionality will differ greatly per company. Some companies may have no multi-plant planning requirements but high subcontracting/contract manufacturing requirements or vice versa. Becoming well versed on superplant requirements and better understanding what is available in software are the first steps to moving towards adaptive supply and production planning capabilities.

CHAPTER 8

The Superplant and the Integration Between ERP and the External Planning System

In this chapter, we will explore issues that relate to the overall integration of multi-plant planning, multi-sourcing and subcontracting/contract manufacturing. Let's begin with an example from multi-plant planning.

Multi-plant Planning Integration Issues

Integration from Galaxy APS to the ERP system is a very important factor to consider when implementing multi-plant planning. In order to implement some of the functionality within the superplant, it is necessary to plan and generate both stock transfers and purchase requisitions, and of course to bring those recommendations over to the

ERP system. In this chapter we will go through the movement of recommendations for all three superplant functionalities.[35]

If we consider the integration between the ERP system and Galaxy APS, for example, Galaxy APS will have routings that span plants; Galaxy APS schedules operations in multiple plants for one manufacturing order.

Multi Plant Planning

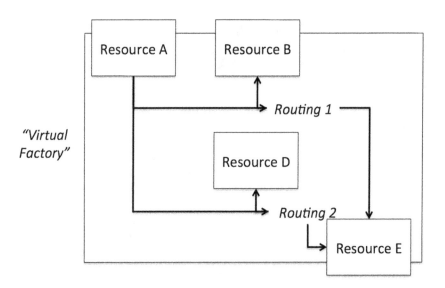

However, the ERP system may not allow selection of the resource and the operation for this when publishing the schedule back to the ERP. To ensure clarity on this point, let us review another graphic.

---

[35] One of the reasons why many companies fail to expand the number of things that they do in their advanced planning systems is because they are hamstrung by their ERP systems. In the SCM Focus Press book *The Real Story Behind ERP: Separating Fact from Fiction,* multiple examples of how many inflexibly designed ERP systems hamper the use of functionality in external applications are provided. This is a great uncovered story in enterprise software because ERP systems have a halo around them for too many reasons to describe here, but it will suffice to say that the costs and limitations they impose on companies is greatly misunderstood.

Multi-plant Integration

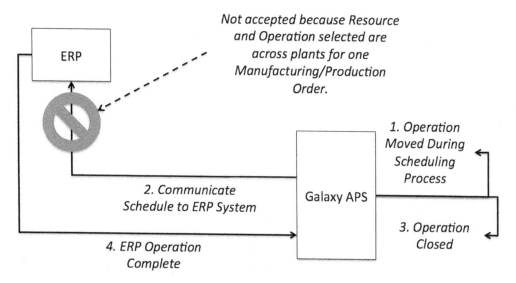

The ERP system may require that all operations are located in one plant. Therefore, adjustments are required to interoperate with an external production and supply planning system that can schedule a single Manufacturing/Production Order across more than one plant.

In this case, just the operation dates could be updated, with the resource information for the operation left alone in the ERP. Also, Galaxy APS might use an Alternate Path with a different set of operations altogether, depending upon which plants or manufacturing methods are used. This would happen if the routings are not kept in sync between the ERP and Galaxy APS. This is less of an issue in an environment with short flow times where backflushing can be used. For those unfamiliar with the term "backflushing," see the following article.

http://www.scmfocus.com/sapplanning/2012/06/28/backflushing-sap/

This same issue applies to subcontracting operations, as the subcontract resource and operation is in an external plant. In both multi-plant planning and subcontracting planning, the routings must be kept synchronized between the systems.

When using multi-plant planning, stock transfers must be created to plan the movement of stock between factories.

Multi-plant Stock Transfers

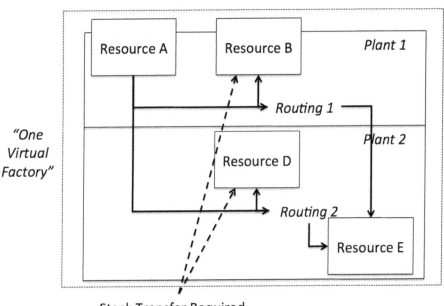

Stock Transfer Required

As has been pointed out, stock transfers can be created either during or after the planning run in Galaxy APS.

Multi-plant Planning Within One Company

Multi-plant planning within a single company is the standard and the easiest multi-plant planning scenario to manage. This simply means that one company controls all of the plants through which the multi-plant routing is "strung." The stock transport functionality in ERP systems generally is designed to move stock through the supply network to final customers, ***neither to reposition stock nor to supply plants in a multi-plant planning environment***. That is, the stock transfer signal must be created by Galaxy APS.

Multi-plant Planning: One Company

Internal Plant 1: USA

Resource A   Resource B   Resource C

Routing 1

Internal Plant 2: USA   *Routing 3*

Resource D   Resource E   Resource F

Routing 2

Here Routing 3 spans Plant 1 and Plant 2. If Resource E has a relatively high capacity—which is not totally consumed by the Galaxy APS Manufacturing Orders within Plant 2—then it can frequently be used for Plant 1 Manufacturing Orders. The stock transfers here are simple in that they are between two plants in the same country.

Multi-plant Planning Within Two Companies: Manufacturing Orders That Span Financial Entities

The second scenario is when the multi-plant planning output creates a manufacturing order that calls on resources with more than *one financial entity.* In one of the test clients where multi-plant planning requirements were compiled, most of the multi-plant planning was between factories in different "companies" (that is the US company and the Croatian company that were both part of the same "parent" company). Components were made in Croatia and then sent to the US. This is a cross-company-code stock transfer, and under SAP ERP such a routing would not be allowed. While there are different companies involved with the planning, this should be transparent to the supply and production planning and scheduling system for reasons of production planning efficiency.

Multi-plant Planning: Two Companies

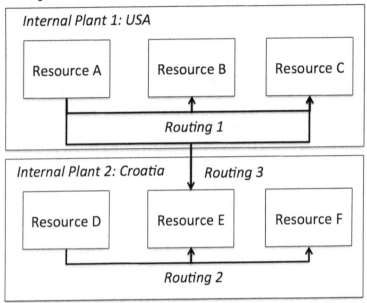

The production/manufacturing order could be kicked off manually in the ERP system for cross-plant routings, and Galaxy APS would be updated as to the status of these operations as part of the normal update (ERP would see the operation as just specific to the plant or as a single plant production/manufacturing order). The problem to be solved is how to trigger the operation in ERP or to implement the multi-plant planning results from Galaxy APS.

Multi-source Planning

With multi-source planning, the integration of stock transfers or purchase requisitions is quite straightforward. The only novel thing that is happening with multi-source planning is that the external supply planning application is choosing among more than one supplier or more than one internal location. However, after the planning decision is made and a stock transfer is created, it simply comes over as a stock transfer between Location A and Location B, instead of Location C and Location B.

Subcontracting and Contract Manufacturing Integration to ERP

With both subcontracting and contract manufacturing planning in Galaxy APS, the output of both of these workflows is a purchase requisition. The purchase requisition is taken into the ERP system quite cleanly because while a subcontractor/contract manufacturing location is treated as an internal location, both are treated as suppliers by the ERP system. This also means that the master data for things like operations and resources for the subcontractor/contract manufacturing location is maintained in Galaxy APS. This is due to the fact that ERP systems do not normally represent manufacturing master data in supplier locations for obvious reasons (these locations are supplier locations to ERP, but planned production locations to Galaxy APS). However, this is quite desirable because Galaxy APS has the ability to very quickly add and manipulate manufacturing master data and can easily be the system of record for subcontractor/contract manufacturer manufacturing master data.[36]

---

[36] ERP vendors or consulting companies will sometimes state in their marketing literature, or verbally on projects, that they are the system of record for all or almost all data. Certainly an ERP software vendor would like to be the system of record for as much data as possible because it increases their control of the ERP system in the client account. This system of record argument is stated in the paper *SAP Business Workflow: The Top 10 Reasons for Using SAP Business Workflow Engine.*

"Your SAP system is your corporate system of record. Doesn't it make sense to have a single system for processing, approving and posting documents?"

Once the ERP system is established in the client's mind as the system of record, then the logic can be used that the other systems, with the superior functionality no less, should not be used, and the inferior functionality in the ERP system should be used. The discussion can then move away from functionality—and towards this nebulous and pseudo-official concept of "system of record." Notice that no evidence has been presented that the ERP system is the system of record for this data—it is simply proposed as a principle.

There is absolutely no reason for the ERP system to be the system of record for all corporate data. Furthermore, ERP simply cannot be the system of record for many types of data because ERP systems *lack the functionality to manage many categories of master data*. A perfect example of this is BOM data. ERP can do almost nothing with the BOM but use it to run MRP and use it for costing for the accounting module. One hundred percent of the time, the BOM begins *its life outside of ERP with design and engineering*. Therefore, the system of record should be the external application, not ERP, as it does not even originate the data. Furthermore, BOM management applications that do originate the data, like Arena Solutions, contain many more BOM management fields than exist in any ERP system. How can an ERP system be a system of record for master data objects for which itself only has a minority of the fields that make up that master data object? It's simple: it can't. Simplistic arguments like this show a great distain for clients and preaches poor data design, and yet they are exceedingly common in the enterprise software market.

Conclusion

Superplant functionalities are advanced; However, they do not present particularly significant integration challenges to the ERP system. Using superplant functionality can mean the locations are modeled differently in the external planning system than in the ERP system, but as long as the recommendations generated by the planning system are brought in as the recommendations from any other external planning system, the ERP system can simply execute the plan without an issue. But of course, as with any external planning system, any client can have nonstandard requirements that will result in custom integration work. PT has a significant amount of integration work behind them with all of the major ERP vendors.

CHAPTER 9

Superplant-enabled Capable-to-promise

One of the exciting positive benefits of superplant planning is how it improves the results of Capable-to-promise (CTP). This combines with the fact that Galaxy APS has CTP functionality that is easy to use. But before we go there, let's start off by reviewing what CTP is.

CTP

CTP is a particular type of checking for availability where the check is performed against the production planning system. CTP is related to available-to-promise (ATP) where the availability check is performed against the supply planning system. How CTP is different from ATP is shown in the following graphic.

CTP Versus ATP

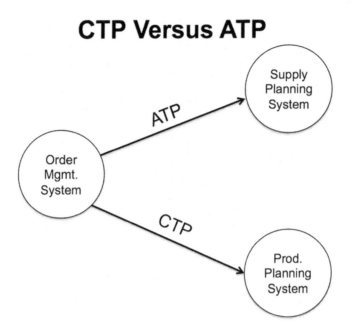

CTP is a check against capacity rather than a check against planned inventory. If capacity is found, the sales order can be accepted and the planned production order scheduled.

CTP could pave the way for a number of companies to meet their stated goals of moving to a make-to-order environment from a make-to-stock environment, meaning that a company could create a production order in response to a sales order rather than to a forecast. The primary reason for companies being unable to move to a make-to-order environment is that the manufacturing and procurement lead times are longer than the lead time between receiving the order and expected delivery of the product. Companies often speak of moving to a make-to-order environment, but often most will not have set up incentive programs to motivate customers to provide their orders further out along the planning horizon.

Secondly, companies are limited in their ability to move to a make-to-order environment in part because they lack CTP functionality.[37] There is a shortage of successful implementations of CTP and this is reflected in the shortage of information published on the topic of CTP. Occasionally CTP is covered in books on

---

[37] Actually there are several that limit most companies from moving to a make-to-order environment; the lack of easily available CTP functionality is just one of these reasons.

planning, and there are a few articles about it although most of them are simply promotional in nature. Upon reviewing the scant literature on CTP, it is apparent that authors who have covered CTP will often write articles on the topic without having performed research into the subject or without having worked on these types of projects. Also, the relationship between CTP and the quality of the production plan is often left out of the discussion. Therefore, it's difficult to gain accurate information about CTP from the publicly available information on the topic.

How CTP Can Be Performed and Integrated with the ERP System

There are several ways to have the CTP results impact the ERP system. Determining which is the most appropriate really depends upon what the company is attempting to accomplish.

1. *Use CTP to get a date but do not create a reservation*: Manually enter the order in ERP. This is the easiest way, as there's no integration of the jobs back to ERP (which can be a challenge in some ERP systems). Since there's no reservation, there's no guarantee that the promise dates will be met once the order is taken—though it could be close.

2. *Make a reservation with CTP then create the orders in the ERP. When they're imported to APS, have them replace the CTP reservation:* This method is a little trickier since there has to be a method of replacing the reservation via integration. It may do a better job at having the actual order scheduled at the same time as the reservation.

3. *Make a reservation with CTP and have the reservation flow back to the ERP system:* This will result in the most accurate execution-to-promise as the reservation is preserved. Integrating the reservation back to the ERP could be a challenge.

4. *Create a quote in ERP and have it schedule in Galaxy APS:* This is the easiest to accomplish since it requires no special integration (since jobs are flowing to Galaxy APS already) or double entry. It lacks the immediate response that is sometimes requested for CTP but perhaps is not really essential (a few hours turnaround may be fine in some situations).

The fourth option is the preferred approach for PT to begin using CTP for the following reasons.

1. It keeps the existing quoting/estimating business process in place, uses the existing data structure, and relies on Galaxy's inherent speed and ease of integration to provide CTP.

2. It also allows planners to move quotes around in the schedule just like any other order, before committing back to Customer Service. It also allows the Galaxy power user to use Galaxy's built-in CTP for quick "what-ifs" if a full quote is not yet done inside the ERP.

However, those companies that really do want to implement CTP via their empowered Customer Service team will want a more real-time response eventually, and this can mean exploring the other options listed previously. This is one reason why SCM Focus recommends an incremental go-live, where simpler functionality is implemented first followed by more advanced functionality. PT follows this methodology as well and refers to it as continuous improvement.

The PT approach to CTP can be described as follows:

1. Commit to competitive and realistic dates (by scheduling before committing).

2. Use Drum-Buffer-Rope and/or other optimization logic to maximize system throughput (and/or other measures).[38]

---

[38] The following quotation on the Drum-Buffer-Rope (DBR) is from the book *Throughput Accounting: A Guide to Constraint Management*. DBR is one of the central concepts to Galaxy APS. *"Comprehending the operational aspects of the theory of constraints requires some understanding of a new set of terms that are not used in traditional company operations. The terms are as follows: Drum: This is the element in a company's operations that prevents the company from producing additional sales. This is the company's constrained capacity resource or bottleneck operation. It will most likely be a machine or person, though it also might be a short supply of materials. Because total company results are constrained by this resource, it beats the cadence for the entire operation—in essence, it is the corporate drum. Buffer: The drum operations must operate at maximum efficiency in order to maximize company sales. However it is subject to the vagaries of upstream problems that impact its rate of production. To avoid this problem, it is necessary to build a buffer of inventory in front of the drum operation. Rope: This term refers to the timed release of raw material into the production process to ensure the job reaches the inventory buffer before the drum operation is scheduled to work on it."* DBR is one of the central concepts to Galaxy APS.

3. Use Buffer Management to manage variability, execute the schedule and deliver on commitments.

Cultural Change as Part of CTP

Many companies don't want Sales making promises without the planner's involvement. While there may be several reasons for this, a common one is that the planner knows more about what's happening in production than Sales and this cross check is beneficial to avoid promises that undermine production efficiency. Therefore, before CTP is rolled out, this change in how a company does things must be undertaken.

Superplant CTP

One benefit of leveraging CTP functionalities within a superplant is that the company can produce more quickly for a given set of resources, suppliers and sub-contractors, since resources are being used more flexibly. CTP can take advantage of this ability to produce more quickly when quoting delivery dates, which can be beneficial in winning additional business.

Performing a CTP Check

Here is the part of the chapter where we will show what a CTP check looks like in the application.

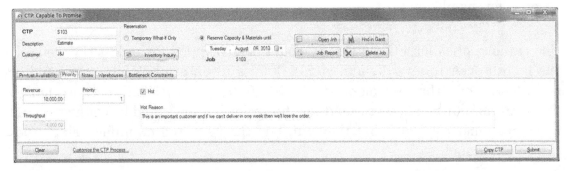

This is the initial order inquiry screen. The request is for the customer J&J, and is a request for capacity and materials to be reserved until August 6 2013. This is one of the options. Another option would be to create a temporary "what if" inquiry. This order inquiry screen also allows for the addition of notes.

We have two products: FG Item 1 and FG Item 2. They have two different need dates. The screen shot below shows how Galaxy provides CTP feedback on these two needs.

FG Item 1 can be completed by 7/1/2013, and is needed by 7/31/2013. So the demand for this product can be met. The only downside is that the product must be kept in inventory. It has a cost of $6,000, which means the company must foot the bill for holding the product in inventory for one month (unless the customer can be convinced to purchase the product early). Most companies use an inventory carrying cost of 25% per year. That works out to roughly 2% per month. 2% x $6,000 is only $120. This is the part about inventory that many "lean" experts don't address; inventory cost can be quantified and in many cases is not expensive to hold.

http://www.scmfocus.com/failedsupplychainconcepts/2009/11/does-lean-make-sense-for-supply-chain/

The same applies for FG Item 2. It would be scheduled for production roughly one month prior to its sale, and at $12,000 its inventory holding cost would be roughly $240 for that one month (2% x $12,000). Notice that in both examples, the result is "On-Time," meaning that this order inquiry could be confirmed and if the customer confirms that they will buy the item, the manufacturing order can be scheduled.

Now we will see what happens when CTP shows that there is no capacity to meet the need date.

Here we can see the same two products, but now FG Item 1 cannot be produced on time because the "ScheduledFinish" date is 7/1/2013, while the "NeedDate" is 6/30/2013. Galaxy APS conveniently calculates how many days the order would be late, which in this case is 1.6 days. When one is dealing with such small differences in the need date and scheduled finish date, it often makes sense to go back to the customer to see if they can wait.

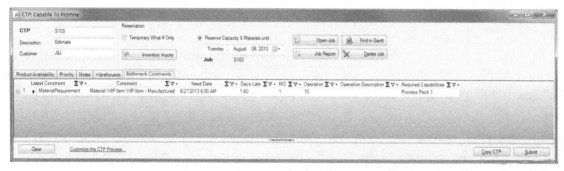

By clicking into the item, one can see the detail on the late product.

These last two screen shots show how CTP can be run in a semi-manual mode. CTP does not accept all orders; it accepts the orders for which there is planned capacity, and moves orders in or out—or even to different plants if available— also based upon planned capacity. Also, CTP keeps production "in charge" in the sense that the planned capacity is simply made available, and then the order can consume capacity.

Here we can see that multiple items can be added to the CTP check just as with a multiple line item sales order. The multiple lines are FG Item 2 for a quantity of 200, FG Item 1 for a quantity of 100 and MRP FG1 for a quantity of 50. Notice as well that the CTP check has been performed and the FG Item 1 has already been scheduled—and is On-Time. For those reading the electronic copy of this book, the Order Job 1 clearly shows as green—meaning it can be scheduled to meet the NeedDate. For those reading the paper versions, one can see the light grey is green, and the dark grey is actually red. If this were not the case the Gantt chart bar would show as red.

CTP can be managed in different ways. It can be automated, or overseen manually or semi-manually. For instance, one could have a production planner/scheduler review all orders that the CTP functionality determined would be "late" and have him or her decide what to do. This addresses one of the major problems with both ATP and CTP projects: it takes too much time to come to a conclusion about the rules which are set up to run them, and most companies are just not comfortable with such an automated process—even though the applications are sold on this functionality. However, in most cases CTP will not be performed manually but will be integrated to the order management system—typically residing in the company's ERP system.

A main advantage of CTP is that it allows a company to partially move away from forecasting. It should be recognized that customers will still want to submit orders within supply/production lead times, which means forecasting will still be required. However, the higher percentage of production that is scheduled with sales orders versus forecasts, the better the utilization of the manufacturing assets. Optimally—and this is really just the optimal state—the company just "makes-to-order." There is a second question as to whether there is inventory below the finished good; in that case it's "assemble-to-order." However, either scenario makes for better asset utilization than a pure make-to-stock environment. Also, the implementing company's customers should be educated to put in their orders as far out as possible rather than waiting—possibly through sharing the benefits through some type of discount. Consider how airlines reduce the price of their "capacity" (airline seats) when tickets are purchased with more lead time. The issue of moving to a make-to-order environment is certainly bigger than simply getting the CTP functionality to work properly.

Historical Background of CTP

CTP has been the "fool's gold" of supply chain planning since it was initially introduced, and it has often failed to gain traction in companies. It is galling to read the scant material on CTP that exists and find that it is uniformly without reference to the problems that CTP has faced in implementation. Software vendors and consulting companies strongly lean towards highlighting categories of software or specific functionalities that have been mildly successful. However, if a category of software or specific functionality has been wildly *unsuccessful*, they simply don't mention the fact. This is essentially lying through exclusion. Instead they say that the software category or specific functionality is a "great idea." This is the opposite of the scientific environment where hypotheses are actually tested, and those that cannot be demonstrated to be true are dispensed with.

The fact that the previous problems with implementations are not analyzed scientifically and implementation methodologies adjusted creates a repetition of implementation problems and is part of the reason that enterprise software does not have a higher success ratio. This is all rather unfortunate because CTP functionality in Galaxy APS works so simply and so well. Secondly, any PT customer receives CTP functionality as part of their Galaxy APS purchase. There is no extra license and once Galaxy APS is implemented, the CTP functionality is operational.

I have given significant thought to the question of why CTP implementations are problematic, and based upon my project experience I have come up with three of the main reasons. There are more reasons than this of course, but I wanted to keep this section to a reasonable length. These reasons were validated with people I have worked with in the past and who have worked in CTP themselves, but I was not able to triangulate them against any publications because literature does not appear to exist on this topic—at least not at the time I performed my literature review for this book.[39]

1. Option Overkill

2. The Length of the Production Planning Horizon

3. Too Much Overhead

Option Overkill

As with more sophisticated available-to-promise (ATP) implementations (such as ATP implementations in the external planning system rather than the basic functionality in ERP systems), CTP implementations suffer from having too many options. For example, if a sales order inquiry comes in to one region, but the region does not have capacity to fulfill the order, should the CTP routine check the factories in the region to the left or right of it? Should it choose a factory that can produce the product a week late but in the first prioritized region, or a week early in a region close by? Are these rules universal, or do they change per region, and per time of year?

Like advanced planning ATP, CTP sounds very nice conceptually, but once one gets into the details of setting the rules, it tends to become quite complicated and often the company itself cannot decide how it wants to set all of the alternatives in the different dimensions. I have been on several advanced planning ATP projects where, after months of discussions and paying for consulting advice, the client decided to simply recreate the elementary ATP logic in SAP ERP; they would worry later about increasing the complexity of the rules so they could take advantage of the advanced ATP functionality in a future release.

---

[39] If any readers disagree with this statement, please add the link to the comment area of this book's web page.

The Length of the Production Planning Horizon

It is standard for the planning horizons to be a year or more. However, on most projects that I have worked on, the production planning horizon is two to three weeks. This is not particularly difficult to change as planning horizons are always adjustable in planning systems; however, if the company has selected a production planning system that includes a time-consuming optimization routine, then extending the production planning horizon may be an issue. This gets to the related topic of the optimizer's efficiency (if in fact optimization is used).

Too Much Overhead

CTP functionality is typically part of an advanced ATP application. This is another application (with its own consultant), which must be integrated both technically and from a process perspective to the ERP system and also to the external production planning system. While it should be simple, some vendors have developed poor designs in this area that overcomplicate this integration. The resulting expense of this tends to limit these types of implementations to the bigger companies. That is unfortunate because smaller companies can benefit from CTP just as much as larger companies. However, even if significant resources are dedicated to the project, the payoff for both advanced ATP applications has been quite poor. It would be difficult to justify implementing a separate module just to perform either ATP or CTP.

Conclusion

CTP can provide a major advantage to companies that implement this functionality because it means that the company can incorporate sales orders directly into the production schedule in real time. It can allow a company to transition to at least partially a make-to-order environment, which is a less wasteful environment.

The first thing to do is to keep CTP expectations and the number of options that are used to a reasonable level. Moving to a full make-to-order environment, with the CTP functionality taking every sales order through every dimension of analysis, is not going to be successful. Very few companies, aside from defense contractors, are truly make-to-order. Furthermore, moving to make-to-order is more than just a function of the available technology, as customers must become acclimated to this capability—and that change will not be immediate. Furthermore many customers prefer to submit their sales orders at the last minute as it maximizes their own flexibility.[40] A more realistic goal is to account for some high-priority scenarios and to use the system to make better-informed order-promising decisions.

The CTP software vendor should have an answer for how they will manage the planning horizon of the production planning system upon which CTP will be based. Responses such as "the horizon can be flexibly configured," do not answer the question sufficiently. Most production planning applications are not designed to have a long production planning horizon. On most Galaxy APS projects, the scheduling planning horizon is roughly three months, and the production planning horizon is often a year. All of this is accomplished with an optimizer running for the full planning horizon, and on hardware that is inexpensive compared to many applications I have seen designed to do the same thing. This "normal" configuration of Galaxy APS means that ***no particular adjustments have to be made to the configuration in order to activate CTP.*** That is, Galaxy APS already has a CTP horizon right from the beginning—and the server that Galaxy APS sits on is already configured for it. It is quite a nice design, and is as close to "out of the box" functionality that one could expect.

---

[40] Some customers play internal politics with their forecast, and a new area of software has developed that helps forecasting departments falsely "improve" their forecast accuracy. This software category is called demand sensing, and it allows the forecasting department to change the forecast ***within the supply lead-time***. To read more about demand sensing see the following article.

> http://www.scmfocus.com/demandplanning/2012/06/is-demand-sensing-being-used-to-fake-forecast-accuracy/

This change has no benefit to the forecasting company, aside from allowing the forecasting department to report inflated forecast accuracy. Another factor which reduces lead time on sales orders is driven by financial KPIs—which are motivated to allow companies to meet quarterly financial objectives. This type of gaming also causes companies to submit their own purchase orders (which are translated to sales orders to their customers) as late as possible. There is often no consideration of the effect on the supplier to such late orders being submitted to them—and if the late order is from a large customer, a salesperson at the supplier will often become involved in order to bring pressure on the manufacturing department to reallocate the capacity away from a customer who has provided ample notice, to the large customer who has given short notice. There is in fact considerable gaming between customers and suppliers and it has worsened in the last several decades, at least in the US. The result is that all of this gaming has quite negative effects on planning.

The vendor should have an answer to the issue of excessive overhead. I have not seen the benefit of having either ATP or CTP contained within a separate application, but I have seen it add too much complexity to client environments. If this is the vendor's approach, how do they intend to mitigate the problems related to ATP/CTP complexity? Galaxy APS does not have a separate application for ATP or CTP. With Galaxy APS, CTP, along with supply planning, production planning and scheduling are all performed from a single application. This means there is no integration to any other system—aside from the ERP to which any Galaxy APS implementation would already be integrated.

CHAPTER 10

Conclusion

Multi-plant planning is the division of what was previously one production process at one location into multiple production processes at multiple locations. Outsourced manufacturing that is to be planned by the OEM works the same way. For software to be said to have superplant functionality, it must be able to choose among alternative routings or paths through multiple locations. Alternate routings can be located inside of a factory or across factories. In traditional production planning applications, the alternative routings that are set up only apply *within a production location*. But for applications with multi-plant planning functionality, the routings apply across production locations, allowing companies to maximize their resources with minimal human intervention into the plan. The multiple routings provide all of the possible permutations that are available within the supply network and the series of factories.

In multi-plant planning, each production process at each location produces to a modular BOM. Any finished goods BOM can be logically subdivided into various sub-BOMs. A sub-BOM is the complete set of parts required to produce a distinct component or subcomponent at a location. When design and engineering create a new BOM (i.e., when a new product is introduced), some of the portions of the BOM may be produced internally while others are purchased. When these components are purchased with no manufacturing planning setup in the OEM's planning system, this is referred to as sourcing and sourcing locations are not modeled as a production location in the planning system(s).

When the OEM that buys a manufactured item from an outsourced entity to which it also provides input materials, this is called subcontracting. With Galaxy APS, subcontractor planning is modeled essentially as an internal location, though not in the ERP system. In Galaxy APS this subcontractor workflow produces a purchase requisition—which is then accepted by the ERP system to eventually be converted into a purchase order. This is required to initiate stock movement from what is modeled as a supplier by the ERP system.

While multi-plant planning is required at some companies, other companies are moving in the opposite direction, away from integrated factories with increased outsourcing of subcomponents where the OEMs essentially take on the role of the general contractor. However as both multi-plant planning and subcontractor/contract manufacturer planning are superplant functionalties, the requirement for superplant-capable software continues to rise.

Any form of outsourced manufacturing, including contract manufacturing, can be accomplished most efficiently by applications that are superplant-capable. Specifically, this means that the application should have the ability to have routings that can string together resources that are located in multiple plants, and that the application has the intelligence to make the appropriate decisions between competing alternatives. These alternatives may be either internal or external production locations. There are many advantages to modeling external production locations in this way, including the greatly underutilized function of performing capacity checking on external production locations.

A superplant planning system can intelligently choose among multiple internal production, or multiple external production alternatives depending upon the circumstance at specific points in time along the planning horizon. The selections among these alternate routings will also change based upon the settings in the

Galaxy APS optimizer. In this way, Galaxy APS can treat all of the factory loca-tions (either internal or external) as if they are logically **one location, i.e. as one "superplant."** Galaxy APS has the functionality to mix and match resources across plants in order to string together the necessary production activities in the optimal sequence—regardless of where in the supply network the resources are located. Because Galaxy APS performs production and supply planning in one planning run and from a single application, there are none of the common discon-nection points that cause inconsistencies between the supply planning system and the production planning system in applications of this type.

Many companies have superplant requirements, but the software they have selected is not capable of planning in this manner. This means that their super-plant requirements simply go unaddressed or must be accounted for with manual planning adjustments. A main reason that this book was written was because no entity has highlighted the importance of these combined functionalities so that companies can recognize that they can address many of their requirements that they currently meet through far less efficient means. For example, superplant func-tionalities are only **some** of the leading capabilities within Galaxy APS. Galaxy APS also contains a highly functional optimization methodology, which combines a duration-based objective function with the ability to adjust the optimizer based upon a wide variety of criteria. PT has not, up until cooperating with this book, actively marketed its superplant functionalities.

Each superplant requirement within a company must be analyzed individually as the need for each superplant functionality will vary quite significantly per company. For instance if a company requires multi-source planning, then a pure supply planning application would be required. But for multi-plant planning and subcontractor/contract manufacturing planning, Galaxy APS contains both these functionalities, and they are surprisingly easy to configure and to put into service. In Chapter 3: "Multi-plant Planning," it was demonstrated how easily the same assignment functionality used to assign Capabilities to Resources or Resources to Capabilities can be used to enable multi-plant planning, as long as at least one of the Resources to be assigned to a Capability was in a different plant. This works the exact same way for planning external manufacturing plants. What this means is that implementing companies can activate some of the most advanced functionality in the supply and production planning software category quickly and inexpensively.

If a company has superplant requirements and has chosen to implement a traditional/sequential supply and production planning system, that implementation is disadvantaged before it begins. Processing production/manufacturing orders that are not locked to a single location provides the system with more flexibility, which will generally mean better planning outcomes and a better utilization of resources. In fact, of all the recent developments in the supply and production planning software categories, SCM Focus regards the adaptive nature of superplant functionalities as some of the most important. While at the time of this book's publication, the superplant is a new concept, software that can manage multi-plant planning, multi-sourcing and subcontractor/contract manufacturing planning has a bright future in the supply and production planning space.

Labor Pools in Galaxy APS

Labor pools are an example of a disconnection point in that they often exist in production planning applications but *not in supply planning applications*. Labor pools exist in Galaxy APS, which as was discussed throughout the book, performs both supply planning and production planning within the same application.

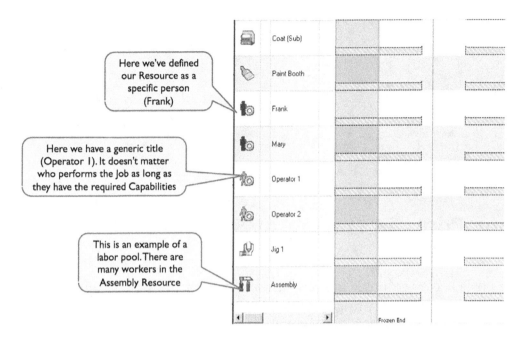

This is the Gantt view within Galaxy APS and is the main user interface for the application. As is shown above, people can be modeled as Resources. This can be a specific person, or a generic operator. As with any other Resource in Galaxy APS, human Resources have Capabilities. When an Operator is used, the Production Rate could be an average of several people, but when a specific person is set up as a Resource, a Production Rate specific to that person's historical productivity, as well as their Work Calendar, can be applied. This calendar works the same way as any other Resource Calendar. At the bottom of the screen we see a Labor Pool Resource, which is assigned to the Assembly Resource.

Labor pools in Galaxy APS are quite straightforward to set up. Galaxy APS uses a Capacity Interval for every resource, and the first step is to set up a Capacity Interval for the labor pool.

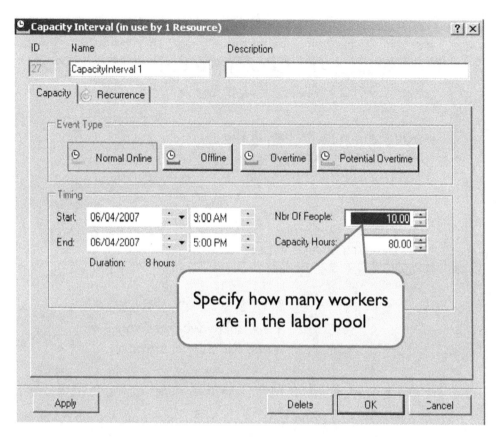

In the Capacity Interval, the following are defined: the number people assigned, the start and end time, and the capacity in hours. These are just the settings in the Normal Online portion of the Capacity Interval. Other areas that can be set up include the Offline, Overtime and Potential Overtime dimensions of the Capacity Interval.

The Capacity Interval can be assigned to multiple resources, and therefore, when the Capacity Interval is updated, it updates all resources for which it is associated.

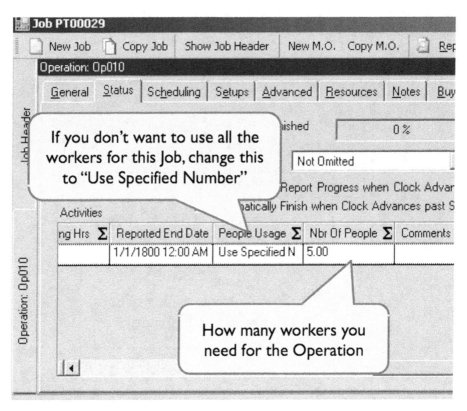

A job can then be assigned to a Labor Pool Resource, where how the job consumes labor can be declared.

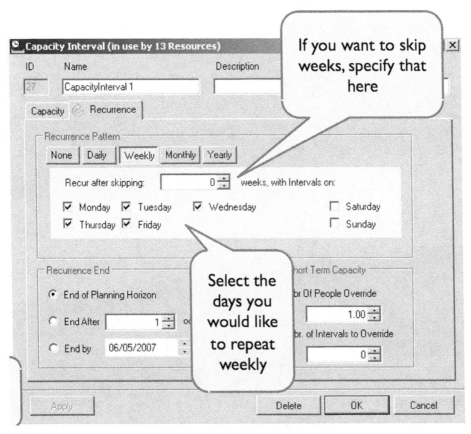

The Capacity Interval can be set to perpetuate or to recur. It can be scheduled to be available for certain days of the week, and to perpetuate for the entire planning horizon or for only a portion of the planning horizon.

Time Horizons in Galaxy APS

As was discussed in Chapter 3: "Multi-plant Planning," Galaxy APS tends to be implemented with a long planning horizon because in effect, Galaxy APS can perform both supply planning and production planning for companies. The planning horizons in Galaxy APS are set in the System Options screen, which is shown on the next page:

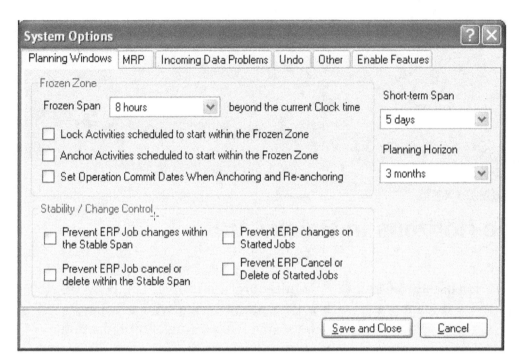

The planning horizons are very simple to set in Galaxy APS in the Planning Window of the System Options. The horizons can be quickly changed here, and when the optimizer is rerun, the new planning horizon is in effect.

The definitions below are all from PlanetTogether's *Advanced Planning and Scheduling Lesson Workbook 6: System Set-up and Security:*

Frozen Span: *Specifies the amount of time from the Clock time that no schedule changes should be made. The purpose of this is to provide stability on the shop floor in the short term, and can (optionally) be used during optimizations to prevent scheduled changes within this time period. The system can be System Set-up & Security 5 setup to avoid changing schedules while leaving flexibility for planners to make manual adjustments. For multi-plant environments you can use each Plant's Stable Span for this purpose (giving each plant the flexibility to have its own fixed horizon).*

Short-Term Span: *Time span from the Clock that constitutes the "short range" schedule. This can be used by various scheduling and statistical functions. It should be set to a value indicating the portion of the schedule (starting from the Clock) on which you want to calculate schedule measures such as utilization.*

Planning Horizon: *Anything beyond the Clock plus this span is considered to be unimportant in the scheduling function. The schedule can extend past this time but optimization algorithms stop detailed planning past this time. All Resources are considered to be continuously available beyond this time.*

Locked: *If an activity is scheduled to start within the Frozen Zone should it be locked automatically to the resource (meaning, it can't be moved to another resource)? It can be unlocked manually if it leaves the Frozen Span.*

Anchored: *If an activity is scheduled to start within the Frozen Zone should it be automatically anchored in time? This does not prevent it from being moved to an alternate resource at the same time.*

Operation Commit Dates: *These dates can be used in reports to compare actual completion dates to scheduled dates for conformance-to-schedule performance measures. Set operation commit dates when anchoring and reanchoring.*

Stability/Change Control: *The Stable Span is defined for each plant and begins at the end of the Frozen Span. It signifies the period of the schedule (in addition to the frozen span) that should be changed as little as possible. It serves as a visual guideline and global optimizations can be set to start outside of this period.*

The Stability/Change Control settings essentially determine how much change can be performed in ERP to manufacturing orders initiated in Galaxy APS. This question of which system is in control of changing the plan is common and constantly arises on both supply planning and production planning projects. Generally, the external planning system has superior user interfaces versus what is available in ERP systems; however, companies typically start from a position of already using their ERP system to make these changes and may want to keep more manual adjustment capability within the ERP system for the sake of continuity.

Changes to the Galaxy APS planning horizon are as easy as making the adjustment to this setting. One could also use the "What If" or simulation functionality in Galaxy APS to make an adjustment to the planning horizon in a simulation version, and run the optimizer to check the results. Galaxy APS also gives the planner the flexibility to use the horizons that are set in the System Options, not only per plant (to process all plants or just a single plant) but also in terms of the horizons.

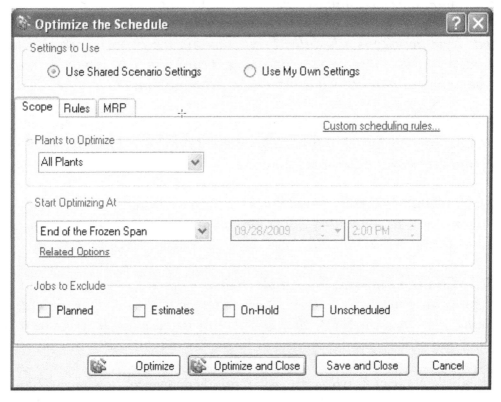

This is the screen that initiates a manually triggered planning run.[41] In the configuration shown above, the optimizer is being instructed to run for All Plants. It can be adjusted to run for just the Plants shown in the Gantt chart (which is the main user interface screen in Galaxy APS) or just for a Selected Plant. The optimizer is told to only start optimizing at the end of the Frozen Span. Obviously, if only one Plant is selected, then

---

[41] Galaxy APS can also be scheduled by a batch job.

no multi-plant or subcontracting/contract manufacturing configuration that has been set up in Galaxy APS will be leveraged.

This also brings up an important issue. Simpler methods can be run for more limited combinations of product and location without negatively impacting the planning output. More sophisticated methods account for more supply and production complexity and if configured properly, will tend to result in superior output. However, with these more sophisticated methods, there is far less ability to run them interactively for small subsets of the problem without negatively impacting other parts of the problem. A perfect example of this is MRP. Because MRP is mathematically simple and cannot account for the interactions of more sophisticated methods, MRP can be constantly rerun at a product-location combination.

Galaxy APS uses its duration-based optimizer along with optimization rules to create the planned production orders.[42] This is used along with MRP in order to create the rest of the supply plan.[43]

---

[42] Galaxy has the option to set cost rather than duration as the objective function. However, no client has ever chosen to use it; they instinctively seem to see the advantages of using time and other factors instead.

[43] PlanetTogether actually provides a few alternatives to its clients. It can simply pass the optimized production plan to the ERP system and allow it to run MRP, or it can run MRP itself and pass the full supply plan and production plan to the ERP system.

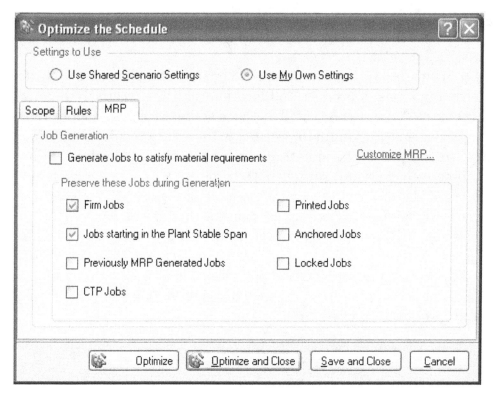

These are the settings for MRP in PlanetTogether.

Prioritizing Internal Demand for Subcomponents over External Demand

How to control the flow of materials between the various sub-plants is one of the major questions with respect to a superplant. When considering alternatives, the method to use for each product location for the finished good must be determined, along with the method to use for the assemblies, components, subcomponents and raw materials.

Companies that use components both for their own finished goods as well as sell components to other companies sometimes want to prioritize the internal demand for components and subcomponents over the demand of external customers. This makes sense from a profitability perspective, as their profit margins on finished goods are higher than that for components. A company may operate a superplant in which component and subcomponent factories supply internal finished goods factories, but where the component and subcomponent factories also supply external customers.

A supply planning method that controls the initial supply and production plan with resource constraints, as well as prioritizes demand types, will meet this requirement. Prioritization and allocation applications such as SAP APO Capable-to-match ("CTM"—an order-by-order method of planning the supply network) are designed to do exactly this.[44]

http://www.scmfocus.com/sapplanning/2009/05/09/ctm/

However, for demand prioritization to work, the implementing company must be able to firmly declare its priority sequence. For instance, it must be able to rank its internal demand above its external customers (or vice versa). In situations where there is no hard and fast rule, prioritization software will not be useful.

In a superplant, demand-type prioritization can sometimes be a requirement. If a company were to allow an external customer to consume capacity for a component for which there is internal demand, and for which they were constrained, then the in-process portions of the finished good would sit in inventory until the component can be produced. This would be an undesirable outcome. When a company wants to prioritize internal demand over external demand, it would make sense to run the allocation supply planning methods with stock transfer requisitions set as a higher priority than sales orders (for instance a "demand" in CTM can be either an external demand or a distribution demand). In this manner, sales orders do not consume components that are needed by internal factories. This is all possible with allocation software.

You can also control which order categories (e.g., Purchase Requisitions, Planned Orders, Sales Orders, etc.) are included in the allocation planning run. In APO, this is set up in the CTM Order Selection transaction. Therefore, CTM provides both the ability to include or exclude order categories, as well as prioritize order categories within a CTM run. Before we get into prioritization, let us cover how order categories are included in CTM to begin with.

---

[44] JDA also offers an allocation application, but I have not used it. There are relatively few applications of this type available in the market.

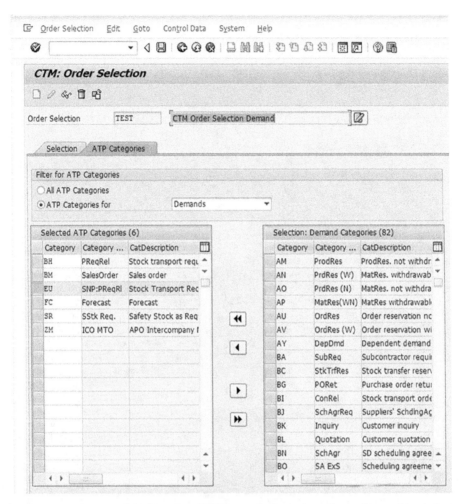

The category groups are added to the CTM run by moving them from the right to the left pane. There are many options, and typically CTM is set up with just a subset of all the order categories that could be included in the run. One can include all the supply and demand order categories into one CTM Order Selection, but generally it is easier if two CTM Order Selections are created: one for demand and one for supply. This provides an extra benefit; with different demand and supply CTM Order Selections, different demand and supply selections can be mixed and matched in various CTM runs.

Notice that I have included "SSTk Req: Safety Stock as Requirement." This is interesting, because safety stock is ordinarily thought of as a supply (just as stock would be considered a supply). However, safety stock is also a demand. That is the safety stock category listed above.

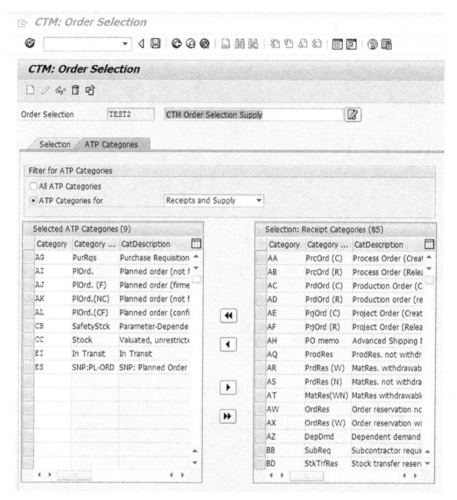

Here is a supply CTM Order Selection that I have created. Again, the sequence is not important; this is the configuration that simply determines which order categories are included and which are excluded from a particular CTM run. After we have included the order categories we want, we can set up their priority. CTM is often discussed as an application that allows for demand prioritization (with customer prioritization as a typical point of emphasis). CTM does much more than this. CTM allows the prioritization of both demand and supply elements; however, it is true that prioritization of demand is a much larger emphasis than prioritization of supply. In most cases the standard prioritization is as follows: planned stock being the first to be consumed, followed by internal manufacturing capacity, followed by external procurement (in the cases where a company both manufactures and buys some part of the product database).

Returning to the topic of safety stock as a demand or supply element, notice that safety stock also appears in the supply order selection. However, here it is "Safety Stock: Parameter-Dependent ATP Safety Stock." This is the safety stock that is available-to-promise, or "planned stock on hand." It would be interesting if safety stock were included in the demand order selection but not the supply order selection, or vice versa.

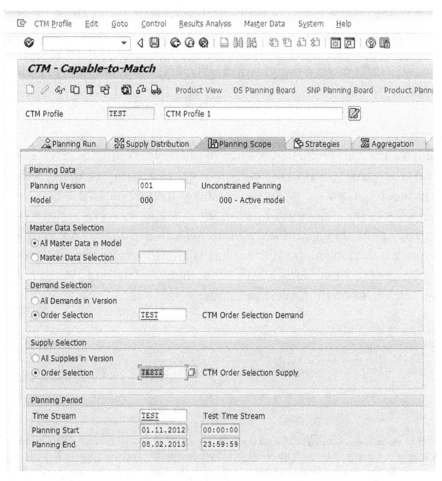

Here is where the CTM Order Selections are assigned to the CTM Profile (see both selections towards the middle of the screen shot). This allows multiple CTM Profiles to be copied over and for them to be assigned to multiple order selections, or any other master data association, such as the master data selection or the time stream shown in this screen shot.

Creating the supply and demand order selections is the easy part of the process. Prioritization of supply and demand is much more difficult because the business must make decisions about how priorities should be set. However, it is quite rare to find an implementing company with a prioritization scheme that can truly be applied consistently. Companies tend to buy the allocation software and fund the project, and in fact get quite far into the project before they understand that they have no consistent prioritization scheme to set up in the system. As described in the following article, this is a common reason for CTM project failures.

http://www.scmfocus.com/sapplanning/2009/12/08/customer-prioritization-and-ctm/

Secondly, the prioritization configuration for CTM is confusing and takes a lot of investment in time and energy to configure properly. While the SNP cost optimizer is also a challenging configuration, considering the more advanced complexity of what the cost optimizer is actually doing underneath the covers, the CTM development team at SAP could have made it a lot more efficient to set up.[45] One simple improvement they could have made was to include both the order selection and order prioritization in a single screen (still allowing separate order selection profiles to be created for demand and for supply). Instead, we get the complex interaction that is shown on the next page:

---

[45] When software runs into repeated difficulties in implementation, this is an indicator that the software itself is at least partially to blame. The constant response from vendors, like SAP, as to why their software is a problematic install is that the users need more training. A common statement by IT is that the business users are to blame. I always find that a strange accusation, as the company itself hired their users—so who is to blame here? Never is the concept aired that possibly the software itself to blame and that the users have already been to training and the system is still not taking because the software is difficult to use—or simply has poor quality output. At SCM Focus we write articles that show how some software is quite difficult to implement and we search for software that implements naturally and easily. There are vast differences in the implementability of software. Companies that would prefer to get the best value from their enterprise software will pay special attention to the actual implementability of the software they purchase.

CTM Profile and Order Categories

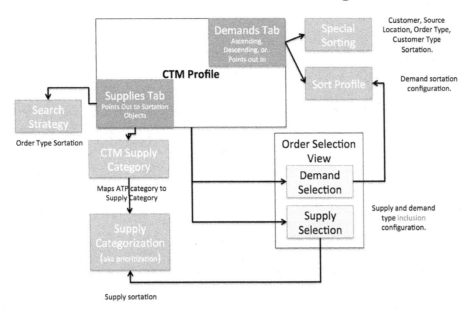

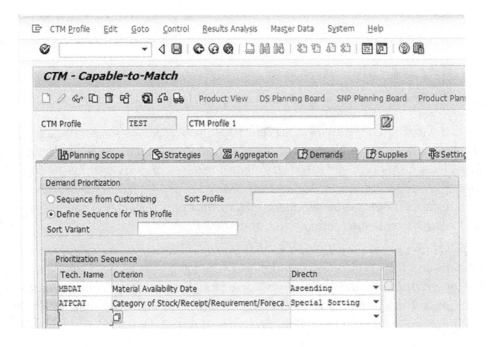

The prioritization of demand starts in the CTM Profile, where each item can be sorted three different ways:

1. *Ascending*

2. *Descending*

3. *Special Sorting*

Ascending and descending can be used for numeric values, such as dates. Because text values don't lend themselves to sorting in this manner, Special Sorting is applied. The Special Sorting points to another configuration screen, where you can choose to apply special sorting to all Stock/Receipt/Requirements/Forecasts in the system.

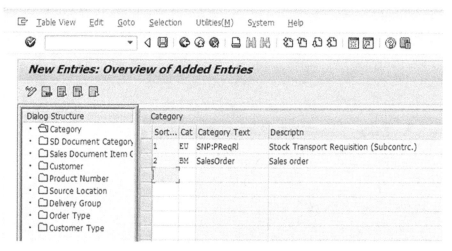

The Special Sorting configuration screen displays categories for sorting. In the above example, I have prioritized Stock Transport Requisitions ahead of Sales Orders. As discussed previously, this is where internal demand receives priority over external customers. Therefore, if a Sales Order and an internal Stock Transport Requisition were competing for the same capacity in a subcomponent factory that was capacity-constrained, the stock transport requisition would receive the capacity/supply.

This configuration is not very difficult to change, so if necessary, the prioritization sequence can be altered as time passes and as the business priorities shift. (I am somewhat conflicted when giving this advice. In companies, business priorities change frequently, but prioritization software is meant for businesses that can construct and articulate their priorities with some clarity and where priorities remain somewhat stable.)

This is a simplified example of a sort profile containing two order categories (sales orders and stock transport requisitions). However, a large number of order categories can be sorted in relation to each other, as is shown in the following screen shot.

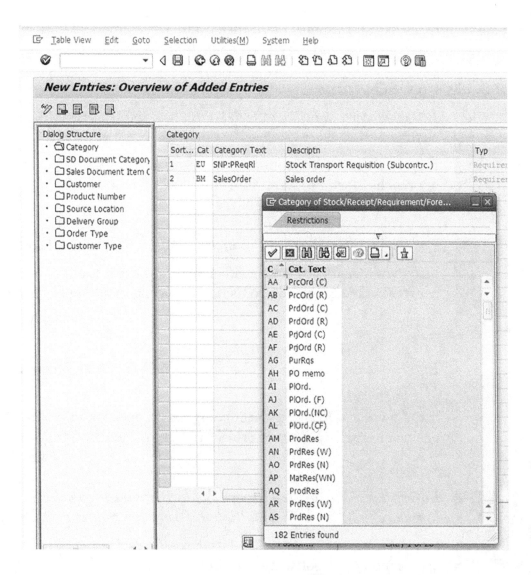

There are many possible supply planning designs for the superplant concept. As previously stated, prioritization may or may not be a requirement. If prioritization is a requirement and if the rules can be properly defined, then an allocation supply planning application can be a good fit. A need to prioritize and a need to plan with constraints are two main reasons that companies with an interest in implementing an SAP product choose SNP and implement CTM. But, prioritization is not necessarily a requirement for every product-location combination. Hence, when developing a superplant supply planning design, some product locations may be planned by CTM and some may be planned by another method.

Other Books from SCM Focus

Bill of Materials in Excel, ERP, Planning and PLM/BMMS Software

http://www.scmfocus.com/scmfocuspress/the-software-approaches-for-improving-your-bill-of-materials-book/

Constrained Supply and Production Planning with SAP APO

http://www.scmfocus.com/scmfocuspress/select-a-book/constrained-supply-and-production-planning-in-sap-apo/

Enterprise Software Risk: Controlling the Main Risk Factors on IT Projects

http://www.scmfocus.com/scmfocuspress/it-decision-making-books/enterprise-software-project-risk-management/

Enterprise Software Selection: How to Pinpoint the Perfect Software Solution using Multiple Information Sources

http://www.scmfocus.com/scmfocuspress/it-decision-making-books/enterprise-software-selection/

Enterprise Software TCO: Calculating and Using Total Cost of Ownership for Decision Making

http://www.scmfocus.com/scmfocuspress/it-decision-making-books/enterprise-software-tco/

Gartner and the Magic Quadrant: A Guide for Buyers, Vendors, Investors

http://www.scmfocus.com/scmfocuspress/it-decision-making-books/gartner-and-the-magic-quadrant/

Inventory Optimization and Multi-Echelon Planning Software

http://www.scmfocus.com/scmfocuspress/supply-books/the-inventory-optimization-and-multi-echelon-software-book/

Multi Method Supply Planning in SAP APO

http://www.scmfocus.com/scmfocuspress/select-a-book/multi-method-supply-planning-in-sap-apo/

Planning Horizons, Calendars and Timings in SAP APO

http://www.scmfocus.com/scmfocuspress/select-a-book/planning-horizons-calendars-and-timings-in-sap-apo/

Process Industry Planning Software: Manufacturing Processes and Software

http://www.scmfocus.com/scmfocuspress/production-books/process-industry-planning/

Replacing Big ERP: Breaking the Big ERP Habit with Best of Breed Applications at a Fraction of the Cost

http://www.scmfocus.com/scmfocuspress/erp-books/replacing-erp/

Setting up the Supply Network in SAP APO

http://www.scmfocus.com/scmfocuspress/select-a-book/setting-up-the-supply-network-in-sap-apo/

SuperPlant: Creating a Nimble Manufacturing Enterprise with Adaptive Planning

http://www.scmfocus.com/scmfocuspress/production-books/the-superplant-concept/

Supply Chain Forecasting Software

http://www.scmfocus.com/scmfocuspress/the-statistical-and-consensus-supply-chain-forecasting-software-book/

Supply Planning with MRP/DRP and APS Software

http://www.scmfocus.com/scmfocuspress/supply-books/the-supply-planning-with-mrpdrp-and-aps-software-book/

The Real Story Behind ERP: Separating Fact from Fiction

http://www.scmfocus.com/scmfocuspress/erp-books/the-real-story-behind-erp/

Spreading the Word

SCM Focus Press is a small publisher. However, we pride ourselves on publishing the unvarnished truth, which most other publishers will not publish. If you feel like you learned something valuable from reading this book, please spread the word by adding a review to our page on Amazon.com.

The Superplant Assessment

In order to advance more companies towards superplant status, SCM Focus and PlanetTogether have joined forces to offer a Superplant Assessment. The Superplant Assessment explains how superplant functionality can be configured and describes the business benefits of rising to superplant status, and assesses how much your company can benefit from using the functionality and your preparedness for such an initiative. To learn more about this assessment, see the following link.

http://www.scmfocus.com/consulting/areas-of-specialty/the-superplant-assessment/

References

Advanced Planning and Scheduling, Lesson Workbook 6: System Set-up and Security. PlanetTogether, Version 2010.1.30.

Beyond BOM 101: Next Generation Bills of Material Management. Arena Solutions, 2011.

Bill of Materials. Wikipedia. Last modified August 31, 2013. http://en.wikipedia.org/wiki/Bill_of_materials.

Bottleneck. Last modified April 24, 2013. http://en.wikipedia.org/wiki/Bottleneck.

Capacity Leveling. http://help.sap.com/saphelp_ewm70/helpdata/en/90/d9733b570d474bba 36bf443f7927c0/content.htm.

Capacity Leveling. http://help.sap.com/saphelp_ewm70/helpdata/en/e3/e9cc13badd4344 a3e0aefb6b9ac265/content.htm.

Chen, Yin-Yann. *An Analytical Framework for Multi-Site Supply Chain Planning Problems.* World Academy of Science, Engineering and Technology, 2010.

Contract Manufacturer. Accessed July 11, 2013. http://en.wikipedia.org/wiki/Contract_manufacturer.

Cost Accounting. Accessed July 17 2013. http://en.wikipedia.org/wiki/Cost_accounting.

Cross Plant Deployment.
http://www.sdn.sap.com/irj/scn/go/portal/prtroot/docs/library/uuid/20049b9
f-e542-2d10-a083-d3cf60dde64e?QuickLink=index&overridelayout=true&
47515223697288.

Cross-plant Routing. SAP Community Network.
http://scn.sap.com/thread/705631.

Cunagin, Cheryl and Stancil, John L. *Cost Accounting: A History of Innovation.*
http://www.ucumberlands.edu/academics/history/files/vol4/CunaginStancil92.html.

Dickersbach, Thomas Jorg and Gerhard Keller. *Production Planning in SAP APO
(2nd Edition).* SAP Press, 2010.

Dickersbach, Thomas Jorg, Gerhard Keller, and Klaus Weihrauch. *Production
Planning and Control with SAP ERP.* SAP Press, 2010.

Fleishmann, Bernhard and Hartmut Stadtler. *Advanced Planning in Supply Chains.*
Springer Press, 2012.

Gaddam, Balaji. *Capable to Match (CTM) with SAP APO.* SAP Press, 2009.

Grischuk, Walt. *Supply Chain Brutalization: The Handbook for Contract
Manufacturing.* BookSurge Publishing, 2010.

Hill, Arthur V. *Encyclopedia of Operations Management: A Field Manual and
Glossary of Operations Management Terms and Concepts.* FT Press, 2011.

Integrating SNP and PP/DS.
http://help.sap.com/saphelp_scm70/helpdata/en/f1/c2d837ffbf2424e10000009
b38f889/content.htm.

Interior Point Method. Last modified August 2, 2013.
http://en.wikipedia.org/wiki/Interior_point_method.

Klier, Thomas and James Rubenstein. *Who Really Made Your Car? Restructuring and
Geographic Change in the Auto Industry.* W.E. Upjohn Institute, 2008.

Lean Planning.
http://events.asug.com/2011AC/1205_Enabling_Lean_Supply_Chain_Planning_
in_SAP_APO.pdf.

McAlpine, Sean. *Cost Accounting is Productivity's Public Enemy Number One.*
Last modified April 9, 2010.
http://www.abonarconsultants.com/blog/2010/04/09/%E2%80%9Ccost-accounting-
is-productivity%E2%80%99s-public-enemy-number-one%E2%80%9D-2/.

Multi Plant Scheduling. PlanetTogether. Last modifed November 18, 2011.
http://www.help.apsportal.com/advanced-topics/multi-plant-and-multi-user-settings/multi-plant-scheduling.

Oboulhas, Tsahat, Xiaofei Xu and Dechen Zhan. Multi-plant purchase co-ordination based on multi-agent system in an ATO environment. Journal of Manufacturing Technology, Vol. 16 Iss: 6, pp.654–669.
http://www.emeraldinsight.com/journals.htm?articleid=1513271.

Okuho, Toshihiro and Tomiura, Eiichi. *Size Matters: Multi-Plant Operation and the Separation of Corporate Headquarters.* The Research Institute of Economy, Trade and Industry: May 2011.
http://www.rieti.go.jp/jp/publications/dp/11e049.pdf.

Opportunity Cost. Accessed June 2 2013.
http://en.wikipedia.org/wiki/Opportunity_cost.

Optimization Based Planning.
http://help.sap.com/saphelp_ewm70/helpdata/en/09/707b37db6bcd66e10000009b38f889/content.htm.

Planning Bottleneck Resources.
http://help.sap.com/saphelp_scm40/helpdata/en/e8/c6765e14e84890b8ce7c0cf7f29384/content.htm.

Planning with Aggregated Resources.
http://help.sap.com/saphelp_ewm70/helpdata/en/43/03b0b3dccd22f3e10000000a1553f7/frameset.htm.

Plant Stability. PlanetTogether. Last modified November 18, 2011.
http://www.help.apsportal.com/advanced-topics/multi-plant-and-multi-user-settings/plant-stability.

Product Failure Rates
http://www.nbcnews.com/id/36005036/ns/business-forbes_com/t/new-improved-failed/
http://drkenhudson.com/why-do-leaders-accept-the-failure-of-a-staggering-percentage-of-new-products/

Pradham, Sandeep and Pavan Verma. *Global Available to Promise with SAP: Functionality and Configuration.* SAP Press, 2011.

Resource Disaggregation.
http://help.sap.com/saphelp_ewm70/helpdata/en/43/03b26bdccd22f3e10000000a1553f7/content.htm.

Rich, Michael. Perpetuating RAND's Tradition of High-Quality Research. November 2011. http://www.rand.org/standards.html.

SAP Business Workflow: *The Top 10 Reasons for Using SAP Business Workflow Engine*. Dolphin Corp. http://www.dolphin-corp.com/top-10-reasons-why-using-sap-business-workflow/.

Schmenner, Roger W. *Multiplant manufacturing strategies among the Fortune 500*. Journal of Operations Management: February 1982. http://www.sciencedirect.com/science/article/pii/0272696382900249.

Second Industrial Revolution. Accessed July 6 2013. http://en.wikipedia.org/wiki/Second_Industrial_Revolution.

Simplex Algorithm. Last modified September 13, 2013. http://en.wikipedia.org/wiki/Simplex_algorithm.

Snapp, Shaun. *Constrained Supply and Production Planning in SAP APO*. SCM Focus Press, 2013.

Snapp, Shaun. *Enterprise Software Selection: How to Pinpoint the Perfect Software Solution Using Multiple Information Sources*. SCM Focus Press, 2013.

Snapp, Shaun. *Multi Method Supply Planning in SAP APO*. SCM Focus Press, 2013.

Snapp, Shaun. *Process Industry Planning Software: Manufacturing Processes and Software*. SCM Focus Press, 2013.

Snapp, Shaun. *Setting Up the Supply Network in SAP APO*. SCM Focus Press, 2013.

Shaun Snapp. *Supply Planning in MRP, DRP and APS Software*. SCM Focus Press, 2013.

Snapp, Shaun. *The Bill of Materials in Excel, ERP, Planning and PLM/BMMS Software*. SCM Focus Press, 2013.

Snapp, Shaun. *The Real Story Behind ERP: Separating Fact from Fiction*. SCM Focus Press, 2013.

Snapp, Shaun. *Planning Horizons, Calendars and Timings in SAP APO*. SCM Focus Press, 2013.

Stock Transfers with PP/DS. http://scn.sap.com/thread/1195661.

Throughput Accounting. Accessed March 23 2013. http://en.wikipedia.org/wiki/Throughput_accounting.

Vertical Integration. Accessed June 26 2013. http://en.wikipedia.org/wiki/Vertical_integration.

Vendor Acknowledgements and Profiles

I have listed brief profiles of each vendor with screen shots included in this book below.

Profiles:

PlanetTogether
PlanetTogether's software is not for supply planning but instead for production planning and scheduling. However, they have been included in this book to demonstrate principles of supply chain optimization. PlanetTogether, software developer of Galaxy APS (Advanced Planning & Scheduling), enables manufacturers to eliminate spreadsheets and connect multi-plant operations for real-time visibility and collaboration. Users report fifty percent reductions in inventory and labor costs, faster time-to-delivery, and a six-month return on investment. PlanetTogether serves nearly one hundred clients in process and discrete manufacturing in a broad range of vertical segments.

www.planettogether.com

SAP

SAP does not need much of an introduction. They are the largest vendor of enterprise software applications for supply chain management. SAP has multiple products that are showcased in this book, including SAP ERP and SAP APO.

www.sap.com

Author Profile

Shaun Snapp is the Founder and Editor of SCM Focus. SCM Focus is one of the largest independent supply chain software analysis and educational sites on the Internet.

After working at several of the largest consulting companies and at i2 Technologies, he became an independent consultant and later started SCM Focus. He maintains a strong interest in comparative software design, and works both in SAP APO, as well as with a variety of best-of-breed supply chain planning vendors. His ongoing relationships with these vendors keep him on the cutting edge of emerging technology.

Primary Sources of Information and Writing Topics

Shaun writes about topics with which he has first-hand experience. These topics range from recovering problematic implementations, to system configuration, to socializing complex software and supply chain concepts in the areas of demand planning, supply planning and production planning.

More broadly, he writes on topics supportive of these applications, which include master data parameter management, integration, analytics, simulation and bill of material management systems. He covers management aspects of enterprise software ranging from software policy to handling consulting partners on SAP projects.

Shaun writes from an implementer's perspective and as a result he focuses on how software is actually used in practice rather than its hypothetical or "pure release note capabilities." Unlike many authors in enterprise software who keep their distance from discussing the realities of software implementation, he writes both on the problems as well as the successes of his software use. This gives him a distinctive voice in the field.

Secondary Sources of Information

In addition to project experience, Shaun's interest in academic literature is a secondary source of information for his books and articles. Intrigued with the historical perspective of supply chain software, much of his writing is influenced by his readings and research into how different categories of supply chain software developed, evolved, and finally became broadly used over time.

Covering the Latest Software Developments

Shaun is focused on supply chain software selections and implementation improvement through writing and consulting, bringing companies some of the newest technologies and methods. Some of the software developments that Shaun showcases at SCM Focus and in books at SCM Focus Press have yet to reach widespread adoption.

Education

Shaun has an undergraduate degree in business from the University of Hawaii, a Masters of Science in Maritime Management from the Maine Maritime Academy and a Masters of Science in Business Logistics from Penn State University. He has taught both logistics and SAP software.

Software Certifications
Shaun has been trained and/or certified in products from i2 Technologies, Servigistics, ToolsGroup and SAP (SD, DP, SNP, SPP, EWM).

Contact
Shaun can be contacted at: shaunsnapp@scmfocus.com www.scmfocus.com/

Abbreviations

(SAP) APO—Advanced Planning and Optimizer

APS—Advanced Planning and Scheduling

ATP—Available to Promise

BOM—Bill of Materials

BMMS—Bill of Materials Management System

(SAP) CTM—Capable to Match

CM—Contract Manufacturer

DC—Distribution Center

(SAP) DP—Demand Planner

(SAP) ERP—Enterprise Resource Planning

(SAP) GATP—Global Available to Promise

RDC—Regional Distribution Center

DBR—Drum – Buffer – Rope

DRP—Distribution Resource Planning

ERP—Enterprise Resource Planning

KPI—Key Performance Indicator

MEIO—Inventory Optimization and Multi-echelon Planning

MPS—Master Production Schedule

MRP—Materials Requirements Planning

OEM—Original Equipment Manufacturer

PDS—Production Data Structure

(SAP) PP—Production Planning

PPM—Production Process Model

(SAP) PP/DS—Production Planning and Detailed Scheduling

PT—PlanetTogether

SKU—Stock Keeping Unit

SCM—Supply Chain Management

(SAP) SNC—Supplier Network Collaboration

(SAP) SNP—Supply Network Planning

S&OP—Sales and Operations Planning

STO—Stock Transport Order

STR—Stock Transport Requisition

(SAP) TLB—Transportation Load Builder

TOC—Theory of Constraints

TP/VS—Transportation Planning and Vehicle Scheduling

VMI—Vendor Managed Inventory

Links Listed in the Book by Chapter

Chapter 1

http://www.scmfocus.com/scmfocuspress

http://help.sap.com/saphelp_45b/helpdata/en/d7/5c9366f47811d1a6
ba0000e83235d4/content.htm

http://help.sap.com/saphelp_40b/helpdata/en/7d/c276fc454011d182
b40000e829fbfe/content.htm

http://help.sap.com/saphelp_46c/helpdata/en/f4/7d2d4d44af11d182
b40000e829fbfe/content.htm

http://www.scmfocus.com/sapplanning/2012/07/26/the-superplant-
concept/

http://www.scmfocus.com/supplychaincollaboration/2013/07/why-
must-specialized-supply-chain-collaboration-applications-exist/

http://www.scmfocus.com/productionplanningandscheduling/
2013/04/22/multi-plant-superplant-planning-definition/

http://www.scmfocus.com/productionplanningandscheduling/

http://www.scmfocus.com/supplychaincollaboration/2010/06/where-
are-the-supply-chain-collaboration-success-stories/

http://www.scmfocus.com/writing-rules/

http://www.scmfocus.com/supplyplanning

http://www.scmfocus.com/productionplanningandscheduling/

http://www.scmfocus.com/sapplanning

http://www.scmfocus.com/scmfocuspress/production-books/the-superplant-concept/

Chapter 2

http://www.scmfocus.com/sapplanning/2012/07/24/synchronizing-integrated-factories-with-stock-transfers/

Chapter 3

http://www.scmfocus.com/scmhistory/2013/08/the-electrification-of-production-plants/

http://www.scmfocus.com/sapplanning/2008/09/21/ppds-and-snp-heuristics/

http://www.scmfocus.com/supplyplanning/2011/07/09/what-is-your-supply-planning-optimizer-optimizing/

http://www.scmfocus.com/inventoryoptimizationmultiechelon/2011/05/socializing-supply-chain-optimization/

http://www.scmfocus.com/demandplanning/2012/03/how-trader-joes-reduces-lumpy-demand/

http://www.scmfocus.com/demandplanning/2012/06/is-demand-sensing-being-used-to-fake-forecast-accuracy/

http://www.scmfocus.com/sapplanning/2009/05/09/ctm/

http://www.scmfocus.com/sapplanning/2009/12/08/customer-prioritization-and-ctm/

http://www.scmfocus.com/supplyplanning/2011/10/02/commonly-used-and-unused-constraints-for-supply-planning/

http://www.scmfocus.com/sapplanning/2012/10/15/understanding-the-flow-of-strs-and-prs-through-apo-with-a-custom-deployment-solution/

http://www.scmfocus.com/sapplanning/2012/10/15/understanding-the-flow-of-strs-and-prs-through-apo-with-a-custom-deployment-solution/

http://www.scmfocus.com/scmhistory/2010/07/the-history-of-apo-and-the-influence-of-i2-technologies/

Chapter 4

http://www.scmfocus.com/sapplanning/2013/07/16/disconnection-points-between-snp-and-ppds/

http://www.scmfocus.com/sapplanning/2012/06/22/firming-in-apo/

Chapter 5

http://www.scmfocus.com/sapplanning/2012/07/06/does-ctm-support-multisourcing/

Chapter 6

http://www.scmfocus.com/productionplanningandscheduling/2013/07/15/subcontracting-supply-chain-definition/

Chapter 8

http://www.scmfocus.com/sapplanning/2012/06/28/backflushing-sap/

Chapter 9

http://www.scmfocus.com/failedsupplychainconcepts/2009/11/does-lean-make-sense-for-supply-chain/

http://www.scmfocus.com/demandplanning/2012/06/is-demand-sensing-being-used-to-fake- forecast-accuracy/

www.ingramcontent.com/pod-product-compliance
Lightning Source LLC
LaVergne TN
LVHW062317060326
832902LV00013B/2273